Pirates of the Florida Coast

Truths, Legends, and Myths

ROBERT JACOB

DocUmeant *Publishing*
244 5th Avenue
Suite G-200
NY, NY 10001
646-233-4366
www.DocUmeantPublishing.com

Published by
DocUmeant Publishing
244 5th Avenue, Suite G-200
NY, NY 10001

646–233-4366

Editor: Philip S. Marks

Assistant Editor: Anne C. Jacob

Map and Woodplate Images provided by Barry Lawrence Rudman Antique Maps Inc. www.Raremaps.com

Ship Illustration pg 41 by: Ginger Marks, www.DocUmeantDesings.com

Cover Design by: Patti Knoles, www.virtualgraphicartsdepartment.com

Layout & Design by: DocUmeant Designs, www.DocUmeantDesings.com

Library of Congress Control Number: 2021040554

First Edition

10 9 8 7 6 5 4

ISBN: 978-1-950075-59-1

DEDICATION

This book is dedicated to the living historians and reenactors throughout the world who devote many thousands of selfless hours in the pursuit of accurately portraying history and keeping it a real and interactive experience for those who attend their events and visit their historic sites.

The vast majority of these devoted individuals do this simply for the love of history, without any monetary compensation whatsoever. Each of them often bears a tremendous financial expense for clothing and equipment which they purchase on their own in order to make their historical interpretation as accurate and real as possible. They keep history alive.

CONTENTS

ILLUSTRATIONS

PREFACE

With each decade, there seems to be one historical period that grabs the imagination and attention of the public. Sometimes this is brought on by the anniversary of a significant historical event—like the bicentennial of the Revolutionary War in 1976. Sometimes the craze is sparked by a blockbuster motion picture. Western reenacting was extremely popular in the late 1990s after the release of "*Tombstone*." Beginning around 2005, there has been a renewed interested in pirates.

From my early childhood, I have been fascinated with different time periods and historical events. Since 1971, I have been heavily involved in historical reenacting and living history interpretation of one time or another from the Renaissance through the late 19th Century. The French and Indian War was my first era. That soon led to the Revolutionary War. *Well, it had to.* The year was 1974 and the entire country was gearing up for the Bicentennial.

Around 2005, my interests shifted to piracy. Not as a practicing participant in robbery on the high seas or illegally copying DVDs, but as a historical interpreter, reenacting the thrilling time of the "Golden Age" of piracy for the general public. For many decades, the bookstores of the world have been flooded with numerous volumes on pirates. But in researching this period, I quickly found that most of the events described in many books written about pirate history were highly contradictory. Also, there is very little written about how they lived, the clothes they wore, the weapons they used, the food they ate, and their often exciting and dangerous lives. Hard facts from contemporary documents concerning pirate history are scarce. After all, as a rule, they were a very secretive group. They seldom filed reports or kept logbooks. There is almost nothing about how they lived or felt about their comrades and their risky, often violent encounters.

But the real problem comes from the fact that the literary market has been flooded with hundreds of books professing to be historically correct—but they aren't! Many modern works about pirates are simply rewrites of older books written many years ago or are based upon local legends and unverified stories. For the books based upon works from the 17th and 18th centuries, it has generally been accepted that those books were historically accurate simply because they were written at a time contemporary to the pirates themselves—but most were not. Many authors of the day were more interested in writing bestsellers than scholarly works of history. For the books based upon local legends, few authors checked the facts. Many of them were content to retell the legends as if they were true. This compelled me to write my first book, *A Pirates Life in the Golden Age of Piracy.* Digging deep into the true history of piracy over a ten-year period, I unearthed a treasure of information that not only tells the truth about these pirates but examines the politics that motivated them. I also delved into their lifestyles as I examined their clothing and fashions, weapons and tactics, food, ships, and attitudes which opened a window into how they lived and what they thought.

I am not a formally trained historian with years of experience researching facts in dusty archives and lonely corridors. I haven't spent dozens of years of my life studying history in higher institutions of learning. I am an officer in the United States Marine Corps who retired after serving for 31 years. But in addition to my military service, for most of my life I have been a historical reenactor; a person dedicated to living history interpretation. I'm not a paid historian, but rather an amateur with a passion for history and historical accuracy. Being so involved with living history has given me a unique perspective on historical events. It has provided me with a first-hand look into the lives of people from the time periods I have recreated.

While selling my first book, *A Pirates Life in the Golden Age of Piracy,* I was frequently asked about pirates in Florida. That makes sense since I lived in Florida and most of my lectures, book signings, and sales events were in that state. Realizing the need for a follow-on book dealing specifically with piracy in Florida, I began examining the historical record and local legends about the pirates and privateers that operated along Florida's 1,350 miles of coastline. But unlike most of the pirates of the 17th and early 18th centuries, where written accounts abound (accurate or not), most of the accounts of Florida's pirates in the late 18th and early 19th centuries seem to primarily come from local legends. These legends are deeply rooted in local tradition and in some cases, have achieved the status of fact in the minds of many people.

Tourism has played a big role in the creation of many of the local pirate legends. Tales of swash-buckling pirates are good for business. Many local legends were invented by tour guides and in some cases fishing guides to amuse and please their clients. The original sources of many of these local pirate stories go back to the beginning of the Florida tourism industry in the late 19th century. Unfortunately for the historical record and for thousands of pirate enthusiasts who seek the truth, many of the late 20th and early 21st century publications about Florida pirates simply retell these invented and unverified local legends or quote older published sources that have done the same thing. In some cases, authors have even invented fictional stories about some of these pirates and represented them as factual.

Association with pirates goes far beyond local legend within the state of Florida. The presence of pirates in Florida has become institutionalized within the state's tourism industry as is evident by the large number of pirate themed festivals. The most prominent one of these is the "Gasparilla Pirate Invasion" held each January and February in Tampa. Additionally, Florida has more pirate themed museums and gift shops than any other state. As if that isn't enough to demonstrate Florida's fascination with pirates, there are many athletic teams throughout the state with names like "Pirates" or "Buccaneers."

I wrote my first book, *A Pirates Life in the Golden Age of Piracy,* to give the reader a comprehensive understanding of the events that transpired during the 17th and early 18th centuries. In *Pirates of the Florida Coast: Truths, Legends, and Myths,* I focus only on those who raided along Florida's coasts. This includes the real pirates and privateers who are said to have visited Florida whether they actually did or not. It also includes the imaginary ones. But in order for the reader to fully understand the proper background and historical perspective of the Florida pirates and privateers, I have chosen to include several chapters and sections from my first book in a capsulized form. For those readers who have not read *A Pirates Life in the Golden Age of Piracy,* this will be vital in building the foundation for a full and complete understanding of the events that led up to Florida's pirates and privateers. For those readers who have read my first book, this will serve as a review of the material that is essential to fully understand the pirates and privateers who operated along the east and west coasts of Florida.

But most importantly, in writing *Pirates of the Florida Coast: Truths, Legends, and Myths,* I hope to separate fact from fiction. Throughout this book, I shall examine the most commonly told stories of famous pirates and privateers who supposedly operated along the Florida coasts and compare local legend to documented fact. In the absence of documented facts, which is quite often the case with many of these legends, I shall examine and discuss the likelihood of truth.

Chapter 1
Let the Piracy Begin

It's a sad but true commentary to say that the first human to build a boat was probably a merchant and the second was probably a pirate. Technically, a pirate is one who commits robbery at sea—well, on the water at any rate. I don't want to limit the definition to just seagoing craft; there certainly have been plenty of river pirates throughout history. However, for the most part, I would like to keep this book focused on the more romanticized pirates and privateers who sailed the seas in the 17th, 18th, and 19th centuries. We begin with the pirates who plundered the Spanish Main in the Caribbean and continue with those who plundered ships and settlements along Florida's coasts.

"The Golden Age of Piracy" is an era that many scholars and authors have identified as approximately the second half of the 17th century to the middle of the 18th century. The exact beginning and ending dates between historians don't always match, but generally this period covers the years from 1620 to 1730. I place the beginning date for my exposé on the "Golden Age of Piracy" at 1640 with the rise of the buccaneers and the ending date at 1722 with the death of the last great pirate, John Roberts, sometimes called Bartholomew Roberts, or just Black Bart. This was when pirates became stylized and our perception of what we think of when we say "pirates" became defined. However, in Florida, this time period was just the beginning. Many of the pirates and privateers who visited Florida did so between the late 18th and early 19th centuries.

In order to fully understand the pirates' role in the political arena, we must first become familiar with the term "letters of marque" (pronounced "mark") and understand the subtleties and differences between a pirate and a privateer. There actually was a huge difference between the two. Their victims wouldn't make any distinction at all, but for the individuals attacking and taking a prize, it meant the difference between acting within the law of their country and acting as criminals. It's the difference between having friendly ports to return to and having to operate in secrecy.

Between 1520 and 1720, it was just too expensive for most nations to maintain large naval fleets during peacetime. They did, of course, have some warships for protection of the major ports and to escort their kings and queens as they traveled abroad, but not enough to effectively wage a naval campaign. When war broke out, governments would have to rapidly expand their small peacetime navies into large wartime fleets. The best way to accomplish this was to hire one. Governments would issue letters of marque to any willing sea captain and crew with a suitable ship, regardless of their nationality, giving them legal permission to attack any nation's ships that were listed in the letters. Those letters were also referred to as commissions. Captains and crewmen sailing under letters of marque were officially called "privateers." The French referred to them as "corsairs." During the "Golden Age of Piracy," a percentage of the profits would be paid back to the government that issued those letters. In England during the 17th century, 1/15th of the profits went to the King and 1/10th to the minister of state or governor who issued the letters or who was appointed to supervise such activity. By the end of the 18th century, this percentage had increased to 50% going to the government. From the mid-16th to the mid-18th century, England, France, and the Dutch Republic all used privateers as part of their official state policy. Between 1702 and 1714 during the War of the Spanish Succession, Queen Anne of England issued 128 privateer commissions through the port of Bristol alone.

Privateers were quite literally naval mercenaries. The ones being attacked always regarded them as pirates, but these men were generally thought of as heroes rather than outlaws by the governments that issued the letters. They were brave naval warriors attacking the vicious enemy and bringing valuable revenue back to the local economy. During the American Revolution (1775–1783) the greatest naval hero of the United States was John Paul Jones. He is considered a hero of the Revolution and the father of the U.S. Navy. But, to the British, he was simply a pirate and was referred to as such. Letters of marque were also issued by heads of state in peacetime. It was a profitable way to wage a cold war against your enemies and generate some revenue at the same time. Governors also issued letters of marque to local privateers as a routine way of protecting their colonies and shipping lanes. As the English, French, and Dutch began colonization in the Caribbean, the use of privateers was an absolute necessity.

Privateer crews were normally a mix of sailors and landsmen. Those lands-men were commonly soldiers, craftsmen, adventurers, shopkeepers, and musicians. Privateer commissions often limited the number of seamen who could be taken aboard to not more than half of the total crew. A privateer captain didn't need a ship full of sailors, he needed fighting men. Privateers only needed enough sailors to handle the ship, the rest were for boarding enemy ships or for attacking cities ashore. For example, in 1719 the English privateer, Captain Shelvocke, only had 20 sailors among his 101-man crew.

Recruiting privateer crews was often done in the same manner as recruiting for any other crew, by public advertising. Notices would be posted around the docks and taverns, coffeehouses, and even in the newspapers. One newspaper ad read, "Captain Peter Lawrence is going a Privateering from Rhode Island in a good Sloop, about 60 Tons, six guns and 90 men for Canada and any Gentlemen of Sailors that are disposed to go shall be kindly entertained."

The first person to raid the Florida coastline was a privateer, Francis Drake. In the late 16th century and throughout most of the 17th century, just about everyone sailing the seas looking for loot was a privateer. Later in the early 19th century, this phenomenon was repeated, when colonies such as Mexico, Venezuela, and Puerto Rico began issuing unlimited letters of marque to anyone in support of their revolt against Spain.

Most of the pirates and privateers who sailed the waters off the coasts of Florida were based in other colonies. English pirates and privateers usually sailed from Jamaica, the Bahamas, or Carolina. French pirates and privateers usually sailed from the Caribbean or from Louisiana. Dutch pirates and privateers usually sailed from the Caribbean. Spanish pirates and privateers sailed from Florida, but they also sailed from the Caribbean. In order to gain the proper background and historical perspective of pirates and privateers who may have visited the Florida shores, it is necessary to fully understand the development of those colonies, especially with regard to the pirates and privateers who lived in those colonies in what many call "The Golden Age of Piracy".

I place the beginning date for the period at 1640 with the rise of the buccaneers and the ending date at 1722 with the death of the last great pirate, John Roberts, sometimes called Bartholomew Roberts, or Black Bart. However, the golden age was not a single period. It was a process of development and growth and remarkable transition and change. As we shall see, the motivation, political support, and social acceptance for these "pirates" dramatically shifted four times during the years that mark the golden age. Some of these changes were brought about by war, some by political treaties, and some were brought about by the riches a country could gain through colonization. Religious freedom and political oppression also played a significant part. Therefore, this golden age can be divided into four distinct subcategories, or eras; 1640–1670 The Buccaneer Privateers; 1670–1702 The Buccaneer Pirates; 1702–1713 The Privateers of Queen Anne's War; 1714–1722 The Pirates.

Chapter 2

The Spanish Main

1500-1630

The Spanish control the Caribbean with well-established colonies in Mexico and in Cuba, Hispaniola, Puerto Rico, Jamaica, Trinidad, and many of the smaller islands.

Hernán Cortés conquered the Aztecs of Mexico in 1521. Francisco Pizarro conquered the Incas of Peru in 1536. Miguel López de Legazpi founded the first Spanish settlement in the Philippines in 1565 and later established Manila as the capital of the Spanish East Indies in 1571.

1570s-1821

Yearly Spanish Treasure Fleets sailed from Havana to Spain

Before we get into all the details of the pirates and privateers who raided or operated along Florida's coasts, we must first understand how the Spanish established colonies in America. Spanish colonization of America began with Christopher Columbus' second voyage in 1494. He claimed the island of Hispaniola for Spain. This island is known today as the Dominican Republic and Haiti. His brother Bartholomew established the capital city of Santo Domingo in 1496 and served as governor. The Treaty of Tordesillas signed by Spain and Portugal in 1497 drew an imaginary line around the world and granted Spain a total monopoly on trade west of that line in the Atlantic Ocean. The Spanish considered the Caribbean to be their own personal sea and North America as their rightful territory.

The capital of the Spanish Caribbean remained Santo Domingo, from 1496 to the early 19th century. The port city of Havana on the island of Cuba was established as the main seaport, where all valuables from the Caribbean would be exported back to Spain. Additional seaports were established to connect their empire. These included Veracruz in Mexico, Maracaibo in Venezuela, and the two ports of Portobelo and Nombre de Dios on the Caribbean side of the Isthmus of Panama. In the Pacific, the primary ports were the City of Panama on the Pacific side of the isthmus and Acapulco along the Mexican coast. By the 1570s, a regular treasure route was established to take the riches from their colonies in the New World and from their colonies in the Pacific back to Spain.

Throughout the year, huge amounts of gold and precious jewels plundered from the Aztecs were sent to Veracruz and then shipped to Santo Domingo. At the same time, valuable commodities from Venezuela and Colombia would also be shipped to Santo Domingo from the port of Maracaibo. On the Pacific side, vast amounts of gold and silver were either taken from the

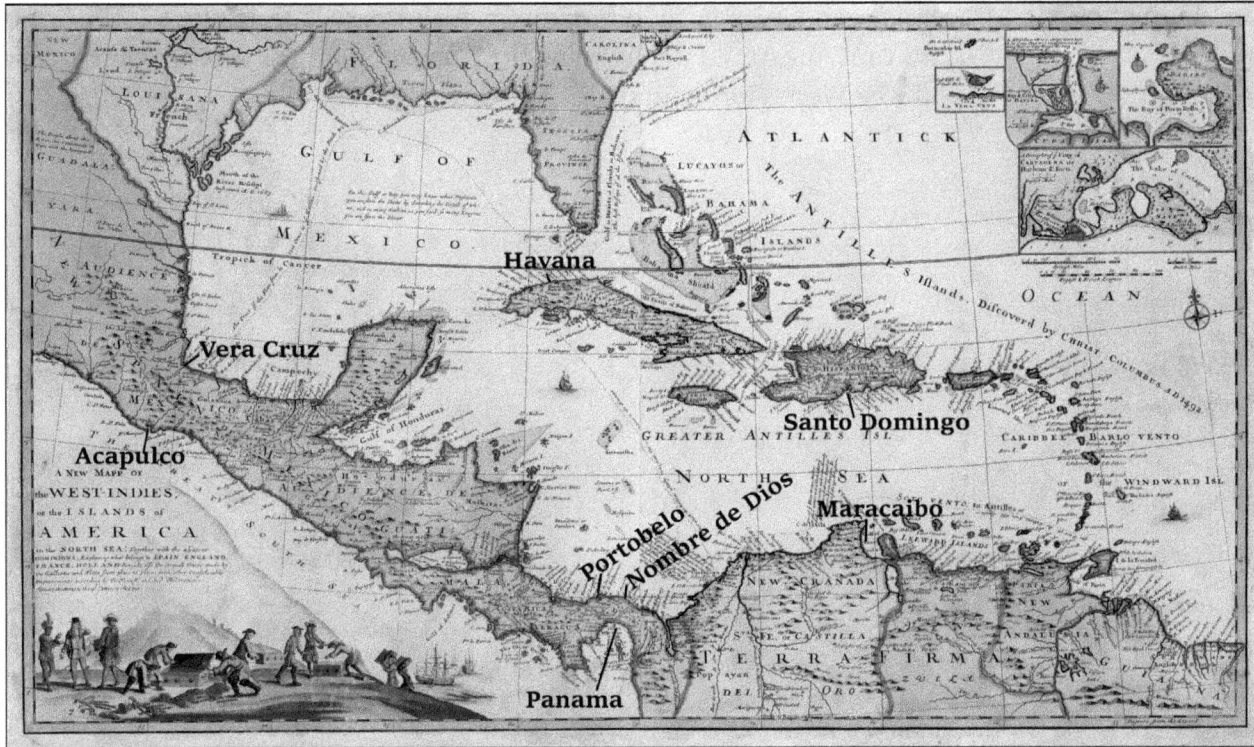

Figure 1: *The Spanish Caribbean*

Incas or mined by slave labor. Those riches would be sent by ship to Panama on a regular basis. Meanwhile, riches such as pearls, spices, and other highly valuable commodities from their colonies in the Philippines were shipped across the Pacific Ocean to Acapulco onboard treasure ships known as Manilla Galleons and then sent on to the City of Panama. All the riches from the Pacific and from South America that accumulated in the City of Panama were then carried across the Isthmus of Panama by well-guarded mule caravans to either Nombre de Dios or Portobelo on the Caribbean side. From there, the riches were shipped to Santo Domingo. As the capital city of the Spanish Caribbean, Santo Domingo received all these riches for processing and accounting before transportation back to Spain. Once all the paperwork was complete, the valuables were sent to Havana where they remained in storage until it was time for the yearly treasure fleet to sail. Once a year, a huge treasure fleet containing all those riches that had been accumulating throughout the year departed Havana and made the perilous journey back to Spain. After leaving Havana, the fleet sailed past the Florida Keys then north through the Florida Straits and along the east coast of Florida. Once reaching the trade winds, the fleet turned eastward and crossed the Atlantic back to Spain.

Because the large treasure fleet was well armed, pirates and privateers were seldom a concern. The real threat came from hurricanes which are common along the Florida coast from May to October. A storm like that could quickly

The Florida Straits connects the Gulf of Mexico to the Atlantic Ocean and is located between Florida (including the Keys) and Cuba and the Bahama Islands. The Gulf Stream is a 75-mile wide current that forms just off of Cancun, and then flows north almost to New Orleans, back down to Havana then up through the Florida Straits all at about the speed of a human walking. It maintains this speed beyond Cape Hatteras, North Carolina, where it veers off to the northeast and eventually dissipates in the North Atlantic. It is a 4,000 mile "river" in the ocean.

destroy an entire fleet and many Spanish treasure ships were lost during this time, the most famous being the *Nuestra Señora de Atocha,* which sank off Key West, Florida in 1622.

Chapter 3
Raiders of The Main

In Europe, Spain was attempting to take over virtually every country possible and had been almost constantly at war with the rest of Europe since 1556 when the Dutch officially allied themselves with England and France against Spain. At sea, Dutch and English "pirates" began redirecting much of Spain's treasure stream to their own governments. The English called themselves "Sea Rovers" while the Dutch referred to themselves as "Sea Beggars." The Spanish called both English and Dutch pirates "Pechelingues." These privateers took prizes wherever they could, in European waters as well as in the Caribbean and the Pacific. Throughout this period, all the waters north of the coast of South America up to Bermuda were referred to as the Spanish Main. These waters included the Caribbean, the Gulf of Mexico, and the waters around the Bahamas. It was called the Spanish Main because the Spanish government mainly considered them as the exclusive property of Spain. As such, the Spanish took a very strict view of any interlopers. In other words, no settlers allowed. Those who tried were met with immediate attack from the Spanish forces which resulted in either their death or enslavement.

The Protestant Reformation was started by Martin Luther in Germany in 1517

The Protestant reformation played a major role in the colonization of the New World, especially for France. Martin Luther began the Protestant Reformation in 1517 and in the 1530s, a French scholar named Jean Calvin began the Calvinist movement, also known as Calvinism. His teachings rapidly spread through Europe. In England, his followers were known as Puritans and in France, his followers were known as Huguenots. While Puritans were accepted in England in the 16th century, the Catholic French didn't take very kindly to this new form of religion.

By the mid-17th century, the Spanish domination of the Caribbean islands and Mexico with all the resources they offered was becoming increasingly annoying to the other European powers. Beginning in 1560, King Charles

IX of France encouraged the Huguenots to form colonies in the New World. To the rulers of France, sending the Huguenots to the Caribbean seemed a "win-win" for all. France got rid of the troublesome Huguenots while establishing a foothold in America, and the Huguenots got to leave France and enjoy religious freedom in their own colonies.

1564

French Huguenots establish a colony near present-day Jacksonville, Florida

Jean Ribault led the first attempt by the French Huguenots to establish a colony in the Spanish territory in the New World. In 1562 his small party built a tiny settlement named Charlesfort on the coast of South Carolina located between the modern cities of Charleston, South Carolina and Savannah, Georgia. Today, the original site of Charlesfort is on the Marine Corps Recruit Depot, Parris Island. However, that first settlement failed due to starvation.

Ribault returned in 1564 and built a larger colony near present-day Jacksonville, Florida. That colony was protected by a star-shaped, wooden fort named Fort Caroline. However, the Spanish discovered the colony and in September 1565 they sent a large military force under command of Admiral Pedro Menéndez de Avilés to destroy it. First, he needed a base of operation with buildings and supplies capable of supporting his force. But instead of a temporary camp, Avilés decided to build a permanent town that could support a new colony. The site he chose was along the shore about 45 miles to the south of Fort Caroline, close enough to launch a land attack. He named this new town St. Augustine.

1565

Spanish forces led by Admiral Pedro Menéndez de Avilés establish the fortified town of St. Augustine, Florida.

Alerted to the Spanish presence, Ribault launched a naval attack, but as his ships approached the Spanish settlement, they were hit by a tropical storm that blew the ships about 15 miles to the south and wrecked them on the beach. While Ribault and his survivors struggled to stay alive on the beach, the Spanish forces marched overland and made a land attack on Fort Caroline. With the majority of the fort's defenders stranded on a beach about 60 miles to the south, the fort quickly fell, and the Spanish destroyed it.

Soon afterwards, the Spanish forces located Ribault and his survivors on the beach near an inlet and executed all of them. The inlet became known as Matanzas Inlet as Matanzas is Spanish for "massacre" or "slaughters". St. Augustine, founded in 1565, flourished as a colony and today is regarded as the oldest city in the United States.

Meanwhile, tension between England and Spain was becoming quite severe. Prior to Queen Elizabeth I assuming the throne in 1558, her sister Mary I ruled England and had married King Philip II of Spain. When Mary died, Philip felt he had a claim on the English throne. Additionally, Elizabeth I was financially and militarily supporting Spain's enemies. England and Spain were now officially political adversaries and war seemed inevitable.

England was desperate for money and taking it from Spain seemed to be achieving both political and financial objectives at the same time. Elizabeth I issued letters of marque to a select group of English privateers with orders to attack Spanish shipping and ports. The most successful of those was Francis Drake. In 1572, Drake successfully landed on the Isthmus of Panama and his men quietly melted into the jungle to wait for the treasure mule caravan that regularly traveled from the City of Panama to the Caribbean. When the caravan reached Drake's men, they successfully ambushed it and took a huge haul of treasure back to their ships and eventually back to England.

Chapter 4
First Privateer Raid on Florida

MAY 27, 1586

Francis Drake was the first privateer in Florida waters, attacking St. Augustine and taking treasure in the amount of about 2,000 gold ducats.

Ducats were gold or silver trade coins with a fixed international value that were often used to conduct business between different nations. In 1586, gold ducats were commonly used in England and the Dutch Republic and had a consistent weight of 3.515 g.

Pinnaces can refer to several different vessels depending upon the nation. In the 16th and 17th centuries, the English used the term "pinnaces" to describe small rowboats carried aboard larger vessels; however, the Dutch used the term to describe much larger two masted sailing vessels.

In 1585, with England threatened by the looming war with Spain and possible invasion by the Spanish armada, Sir Francis Drake seemed like England's best hope. Drake returned to the Caribbean late in 1585 with a fleet of 21 ships. On January 1, 1586, he made an amphibious assault taking the Spanish capital city of Santo Domingo, plundering the town, and leaving with 25,000 ducats in treasure. A month later, Drake repeated his actions by taking and plundering Cartagena, leaving with an additional 110,000 ducats in treasure. From there, he sailed north through the Florida Straits and along the coast until he arrived at St. Augustine. It was May 26, 1586, when Drake's 21 vessels suddenly appeared at the entrance to the harbor.

The following day, May 27, 1586, Drake ordered approximately two hundred of his men to climb over the side of their ships and into pinnaces which had been lowered from his ships earlier that morning. With no artillery, Drake's well-armed men rowed their small boats up the inlet and directly towards a Spanish log stockade named Fort San Juan. This fortification mounted fourteen bronze artillery pieces and guarded the entrance to the town. After the Spanish fired a few ineffective shots, the English landed and swarmed over the walls. By then, the Spanish had fled, leaving the fort deserted. Fort San Juan had been captured with only a few casualties. But in their haste to run from the English, the Spanish had failed to destroy any of their artillery pieces. They also left behind a chest containing about 2,000 gold ducats which was intended to be used as the garrison's payroll. The gold was quickly loaded onto the pinnaces and the privateers proceeded towards the main settlement of St. Augustine itself.

Upon arrival, Drake's men found no sign of any Spanish soldiers. In fact, the town was totally abandoned. As Drake's men wandered through the empty streets and buildings, they found almost nothing of value. But as the

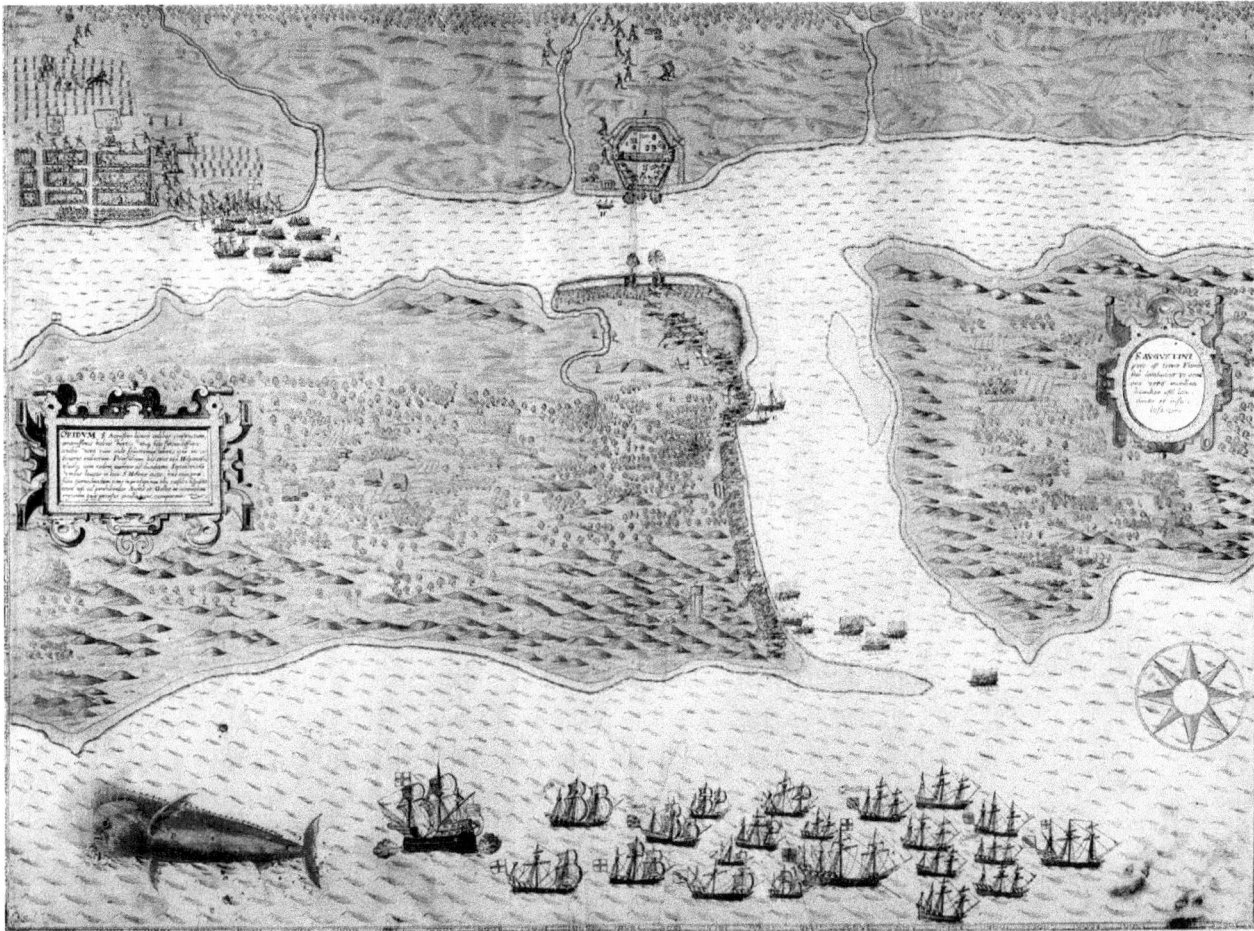

Figure 2: *St. Augustine 1586*

privateers reached the outskirts of the town, they were suddenly confronted by a skirmish line of Spanish soldiers. Suddenly, a thunderous roar of gunfire came from the Spanish line and bullets whizzed past Drake's men. A few of the bullets found their mark and several of the privateers were either dead or wounded. Drake's men quickly rallied and returned fire. Once again, the Spanish fled. The English privateers remained in the town overnight searching for valuables.

The following day, Drake destroyed St. Augustine. All the buildings and crops were torched and anything of value that Drake's men found was either taken or destroyed. Traveling back to their ships, Drake's men also destroyed Fort San Juan and took the 14 artillery pieces. The only substantial profits of the entire operation were the 2,000 gold ducats they found at the fort and the 14 artillery pieces. Drake's real success came from the destruction of a Spanish city in the New World and the political statement that it made.

Chapter 5
Chipping Away at The Main

1607

First permanent English settlement in the New World is established at Jamestown, Virginia

Throughout the 16th century, the French and English had made several attempts to establish settlements in North America, but they all failed. This changed in the 17th century. In 1607, the English finally established their first successful colony in the New World. Jamestown, named for King James I, was not just a settlement, but was strategically situated far enough away from the Spanish as not to draw their attention but close enough to their shipping lanes to serve as a base of naval operations if England wished to attack their fleets. Other English settlements from Virginia to Maine quickly followed. But there still weren't any English, Dutch, or French settlements in the Caribbean.

1618–1648

Thirty Years War in Europe begins over religious persecution

The situation for Huguenots and other Protestant denominations quickly deteriorated in Europe with the beginning of the Thirty Years War (1618–1648). Originally started over political control of Bohemia, the war rapidly transformed into a war of religious persecution. In the 1620s the Huguenots began looking for places to settle in the Caribbean. As the 17th century progressed, King Louis XIV's Catholic government in France went from legal persecution to full-scale slaughter in order to rid France of the Huguenots. Even those in Europe who weren't persecuted for their beliefs were anxious to get away from the turmoil of a vicious war and for many the Caribbean seemed like a good refuge.

The Lesser Antilles islands at the eastern end of the Caribbean appeared to be the safest location. Those islands were out of the main Spanish shipping lanes and far from the large fortifications on Cuba. English settlers began arriving in 1623, when Sir Thomas Warner established the first British colony in the West Indies on St. Kitts. The French government was on good terms with England during this period and wanted to get in on the action too. In 1625, the English willingly partitioned St. Kitts giving half of the island to the French. This became the first French colony in the Caribbean. Other

English colonies soon followed—Barbados in 1627, Nevis in 1628, and soon afterwards, Antigua, Montserrat, Anguilla, and Tortola. Meanwhile, the French established settlements on Martinique, the Guadeloupe Archipelago, and St. Barts. The Dutch also wanted to get in on the action and established a small colony on St. Maarten's (St. Martin's) Island where they began a modest salt mining operation. Later on, in 1634, the Dutch colonized the island of Curaçao which became the Dutch center of operation in the Caribbean. The Spanish attempted to stop this colonization, but every time they attacked a settlement, the settlers would disperse until the Spanish left.

Figure 3: *First Settlements*

Around 1625, some of the French Huguenots and English adventurers who had recently arrived on St. Kitts, relocated westward and established a small settlement on a tiny island off the northwest coast of Hispaniola. The island's name is Tortuga, and it is 12 miles long and 6 miles wide. The island was originally inhabited by the Tainos. First named Santa Ana by Columbus in 1493, the 17th century French named it Ile de La Tortue meaning Turtle Isle. This name eventually became Tortuga. The new settlers built a small seaport and town named Cayenne along the natural harbor on the south side of the island. Almost immediately after Cayenne was established, the English and French settlers began to migrate to the rich interior of Hispaniola. Tortuga was close enough to the north shore of Hispaniola for dugout canoes to be used to travel back and forth.

Located to the east of Cuba, the island of Hispaniola is among the largest islands in the Caribbean. Today, it contains the nations of the Dominican

1620s AND 1630s

French, English, and Dutch begin small settlements in the Caribbean on Barbados, Hispaniola, Nevis, St. Kitts, Martinique, Guadeloupe, Tortuga, and Curaçao.

Republic and Haiti. It has rich and fertile soil, lush vegetation, and an ideal climate well-suited for agriculture. Named "La Espanola" or Spanish Island by Christopher Columbus, the island is where his famous ship, the *Santa Maria*, ran aground and sank. The capital city of Santo Domingo on the south side of the island dates back to 1496 and was officially established as the capital on August 5, 1498. It was the center of Spanish political authority in the Caribbean from the 16th to the early 19th century.

Thousands of Spanish settlers, plantation owners, and bureaucrats flocked to Hispaniola throughout the 16th century, bringing a large population of slaves with them. However, they primarily settled on the eastern end where the farming was good or along the south coast close to the main shipping port and capital city of Santo Domingo. This left the vast interior, the western end, and the northern coast completely unoccupied.

Even though Santo Domingo was the central point of the Spanish government in the Caribbean, the north side of Hispaniola had no Spanish settlements and seemed like an ideal place for the French and English settlers to establish a foothold in the Caribbean. Unlike the settlers on Nevis and St. Kitts, they were sharing their island with their enemy, the Spanish. Consequently, the settlers on Hispaniola took every precaution not to be seen.

Figure 4: *Hispaniola*

Chapter 6
From Pork to Gunpowder

Living under constant threat of Spanish attack, the French and English settlers on Tortuga and Hispaniola had developed a uniquely different type of lifestyle. They had to remain extremely mobile just in case the Spanish discovered them. If they were found, they would have to get away quickly by disappearing into the vast woodlands. Farming of any sort was totally out of the question because it took a lot of land, a lot of time, and could not be hidden. Large crops would be fairly easy for the Spanish to find and destroy. The settlers needed to find another means of making a living, and they found it.

For over 100 years, livestock had been escaping from the Spanish settlements to the south and breeding in the remote forests. The woodland of the interior was filled with cattle and pigs. It was a hunter's paradise. Hispaniola, like many of the islands in the Caribbean, was originally inhabited by the Tainos, an Arawak-speaking people, believed to be originally from South America. Most of the Tainos had been killed by the Spanish or died from disease from European contact; however, there were a few left when the English and French settlers arrived. The Tainos had a very unique way of preparing meat. They wrapped the meat in wet leaves and placed it on a wooden platform over a very smoky, slow burning fire. In the Arawak language of the Tainos, that process was called "boucan" and the Arawak word for the dried meat it produced was "charqui," which was pronounced "jerky" in English. Today, we call this process "pulled" or "jerked" pork or beef. The settlers perfected this technique that they had learned from the Tainos and went into the meat processing business. Hence, those early hunters and cookers of meat were called "boucanniors" by the French and "boucaneers" by the English.

The French boucanniors and the English boucaneers worked together, hunting wild boar and cattle deep in the interior of Hispaniola, all the while

Mid-1620s

Early English and French hunters settle on Hispaniola.

evading the Spanish patrols. To sell their product, they established coastal trading settlements and small ports on the western part of the island away from the Spanish sphere of influence. The Spanish continually attempted to capture or kill these interlopers. Occasionally, the Spanish were successful, but generally the boucanniors or boucaneers would simply disappear into the jungle and return after the Spanish left. Since their operations were illegal in the eyes of the Spanish authority, their customers tended to be smugglers and other settlers also operating outside of Spanish control. They would trade for the necessities of life including weapons, powder, shot, and of course, wine.

Their appearance would have been more like that of mountain men from the American west during the 1820s. They were hunters and dressed as such. Contemporary accounts describe them as wearing coarse shirts and other clothing made mostly of animal skins. No fancy garments, just homemade clothing that is adapted and practical for travel in the jungle. This lifestyle required them to be heavily armed with muskets and assorted skinning knives. They had to be prepared to both hunt wild boar and defend themselves against Spanish patrols. Additionally, they probably smelled horrible. Anyone who has spent time near a meat packing plant would testify to that. According to a French clergyman, Abbe Jean Baptiste Du Tertre, "these are the butcher's vilest servants who have been eight days in the slaughterhouse without washing themselves."

By 1629, the Spanish had had enough. Not only were these settlers from enemy nations, but to the Catholic Spanish, they were heretics too. These English, French, and Dutch invaders were all predominantly Protestants. The task of driving these heretics out was given to one of Spain's ablest generals in the 30 Years War, Don Fadrique de Toledo. It was a Spanish "maximum effort" to rid the Caribbean of the heretics and foreigners once and for all. His first targets were the English and French settlements on the islands of Nevis and St. Kitts. He launched major military assaults on both colonies in 1629 and captured many of the settlers, but just as many managed to escape into the interior of the islands they were on or fled to nearby islands. As soon as the Spanish left however, the escaped settlers simply returned and rebuilt their settlements. But many others decided to seek refuge and safety on the tiny island of Tortuga, just off the northern coast of Hispaniola, which seemed to be a fairly defensible position.

Even though Don Fadrique de Toledo's forces enjoyed some immediate success destroying settlements and rounding up prisoners, there was no permanence to his efforts. As soon as the Spanish forces left, the English, French, and Dutch settlers who escaped would return and rebuild. Overall, his campaign was a failure. Don Fadrique de Toledo did manage to achieve one thing, however. His attacks instilled a deep hatred for the Spanish among all the settlers. This hatred united them against the Spanish for the rest of the century.

But Don Fadrique's attacks did something else. They forced the French boucanniors and the English boucaneers to seek refuge on Tortuga. In the four years prior to the Spanish attacks of 1629, the French and English settlers had established a small port town called Cayenne on a natural harbor on the south side of the island. Although it probably was unimpressive at the time, it had a market that was used primarily to sell beef and pork. It was also a central trading spot for smugglers. But beginning in 1630, with Tortuga as their stronghold, the French boucanniors and the English boucaneers became organized into a resistance force. They decided to use their hunting skills against men instead of animals. In a maximum effort to defend themselves, they completed their fortifications and made Tortuga their center of resistance and operations.

Tortuga began a rapid development into a major port with impressive fortifications. Today, it is the most famous pirate port of all time and has reached a legendary status within pirate lore, stories, novels, and Hollywood films. Despite this status, it's a real place and very deserving of its reputation. The harbor of Cayenne on Tortuga was very close to the main Spanish shipping lanes making it fairly easy to prey on small Spanish ships. The French boucanniors and the English boucaneers continued their resistance against the Spanish. Since offense is the best defense, they began taking the fight to the Spanish instead of just waiting for an attack to come. The more they fought back against the Spanish, the more they realized that taking Spanish property was far more profitable than selling meat to the odd passerby. I'm sure that a certain amount of revenge for previous Spanish atrocities also played into the situation. Additionally, capturing Spanish shipping was undoubtedly a lot more fun.

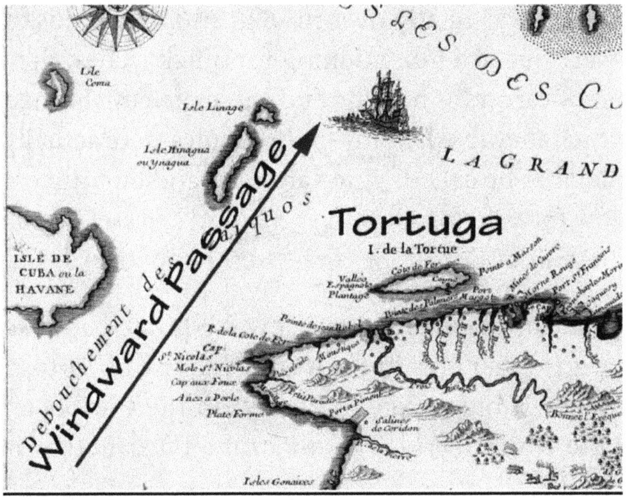

1631

Tortuga is established as a stronghold against Spanish attacks

Figure 5: *Tortuga*

1640-1653

Jean le Vasseur was Governor of Tortuga

Soon, Dutch adventurers began arriving to join the English and French. Their strength grew significantly and they began a French/Dutch/English partnership against the Spanish that would dominate the Caribbean for the next 60 years. The hunters, processors, and sellers of pulled pork made the transition to a warrior culture that took ships at sea. French boucanniors and the English boucaneers became known as buccaneers and forever afterwards, the word "buccaneer" would be synonymous with "pirate."

The first era of the "Golden Age of Piracy" had begun. It was the era of the buccaneer privateers. The buccaneers on Tortuga were well supported by the English, French, and Dutch governments who wanted to keep the Spanish away from their colonies. Letters of marque were issued that legitimized their actions. France finally took control of Tortuga and officially appointed Jean le Vasseur as the first Royal French Governor of Tortuga in 1640. He was an engineer as well as a politician and immediately began construction of a large and imposing stone fortress with 40 guns on a rocky hill overlooking the natural harbor. The fort was named Fort de Rocher and it commanded the entire city. Le Vasseur also issued letters of marque against the Spanish to all who applied. This act alone served to encourage buccaneers of all nations to flock to Tortuga. If that wasn't enough, in 1645 he imported 1,650 prostitutes to Tortuga in an attempt to bring harmony and control to the unruly pirates. At least that was his official story. From 1645 through the 1660s, Tortuga lived up to every elaborate description and portrayal in novels and in Hollywood films. It was a seaport filled with taverns, brothels, smugglers, and buccaneers. It was a no holds barred kind of town with a huge fortress on top of a hill to protect the pirates against the Spanish. It was a place where buccaneers could walk the streets and trade all their stolen booty for whatever they wanted free from worry of being arrested. It was a port with large buccaneer fleets in the harbor refitting for their next venture.

1648

Thirty Years War ends with the treaties of Osnabruck and Munster

In Europe, the Thirty Years War was finally over in 1648 and thousands of soldiers and mercenaries were out of a job. Looking for quick riches, they flocked to Tortuga to begin new careers as buccaneers. These men established rules and conventions by which they lived. Many of these rules were actually written down in what would later be called "ship's articles" and buccaneers would have to agree to follow them before signing on board a vessel. It was this time when these buccaneers developed into the "Brethren of the Coast."

By 1650, Tortuga had become a buccaneer's haven. It was a pirate paradise that was totally unique from any other port in the world. The rise of Tortuga and the buccaneers launched a "Golden Age of Piracy" in the Caribbean which spread throughout the entire world. The era of the buccaneer privateers had begun.

Chapter 7
Port Royal, Jamaica
The New Buccaneer Center

Taken by a large invasion force in 1655, Jamaica emerged as the central English colony in the Caribbean. The English were very quick in establishing their new colony and built a large seaport with three well-built forts on a sandbar. Originally, that port was named the Port of Cagway. However, fear of a Spanish counterattack was always present. To defend Jamaica, Captain Christopher Myngs of the English navy organized the buccaneers from Tortuga into large buccaneer fleets and went on the offensive, attacking the Spanish throughout the Caribbean. Jamaica soon emerged as the center of all buccaneer activity in the Caribbean.

The focus of these fleets were large-scale amphibious operations against Spanish ports. Instead of taking ships at sea, these buccaneers would land ashore near a major port and take it from the land using infantry tactics. Many of these fleets were well sponsored by the governments they served. Privateers were used in the Caribbean as the regular standing navy. The privateers waged large-scale war against shipping and trade, all on the orders of the crown. Their ships varied greatly in size and composition, but it was not unusual to see large buccaneer warships of 30 to 40 guns.

In 1660, the English parliamentary government known as the Commonwealth was deposed and the royal family was brought back from exile. King Charles II was restored to the throne in a time period known as "the restoration." In Jamaica, the Port of Cagway was renamed Port Royal in honor of the king. Additionally, the political leaders in the Caribbean who had been supporters of the Parliamentary government began to be replaced with those who had remained loyal to the king. But Captain Myngs didn't care about one's politics and used everyone he could get.

MAY 1655

An English invasion force of 7,000 troops and 30 vessels under the command of Robert Venables and William Penn invade Jamaica.

1660

King Charles II was restored to the throne in a time period known as "the restoration."

1660

Port Royal emerges as the new buccaneer haven.

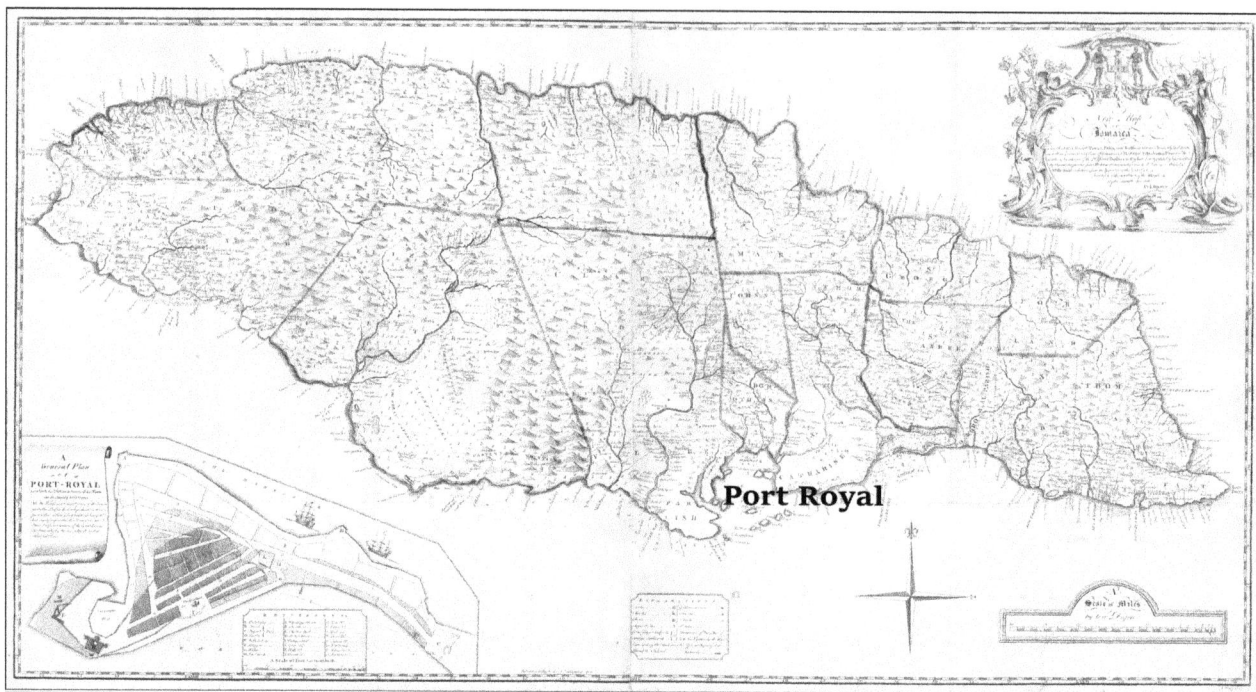

Figure 6: *Jamaica*

Jamaica's main port of Port Royal emerged as the new buccaneer center of the Caribbean. In the 1660s it acquired a reputation of being "The Sodom of the New World" and "The Dunghill of the Universe." Actually, Port Royal really wasn't such a bad place. Most of the negative descriptions of Port Royal seem to have come from people who have a dislike for prostitutes, drinking, and gambling. I rather think of Port Royal as the Las Vegas of the day. By 1660 Port Royal had become the New World's first real boom town. It was like San Francisco during the 1849 gold rush or the city of Las Vegas in the late 20th century. It was a place where anybody could earn a fortune in a short time. There were over 6,000 local residents and one of every five buildings was a brothel, gambling house, or tavern. Thanks to the buccaneers, Port Royal had become the most important English port in the Americas, a mercantile center of the English Caribbean, with vast amounts of goods flowing in and out of its harbor as part of an expansive trade network. Port Royal was an ideal place for privateers to sell their captured cargo and to spend their money.

The population was extremely diverse. In addition to the English Anglicans, there were accounts of merchants who were Quakers, Papists, Puritans, Presbyterians, and Jews. In fact, the first synagogue built in the New World was in Port Royal. All those diverse groups practiced their religion openly alongside the free-wheeling buccaneers who frequented the port. Additionally, they all had a friendly "Welcome to Port Royal" attitude towards buccaneers.

Buccaneer crews consisted of mostly soldiers with a few sailors to handle the ship. But some of the buccaneers were even respected shop keepers, business owners, and craftsmen themselves. A tradesman could make a year's salary in just one month sailing as a member of a privateer crew. Their appearance would have been a mix of fancy clothing taken from prize vessels, nautical clothing used onboard as a matter of practicality, and contemporary military uniforms with the customary military boots. However, leather sandals were also worn. By the late 1660s, buccaneers looked more like land soldiers than sailors. With each raid, the captured clothing contributed greatly to their eclectic appearance. They all would have been well-armed and provisioned when conducting a raid and also when on shore leave. The fact that they were exceptionally well-armed ashore is the first thing many contemporary observers mention. Even though the towns had some sort of law, they didn't have a police force. It was every man for himself when they went ashore. Contemporary accounts mention a buccaneer named Rock Brasiliano who would "scour the streets of Port Royal with a sword, dismembering anyone who crossed his path." If a buccaneer wanted to keep his money, he had to be prepared to fight for it.

In the early 1660s, Myngs augmented his buccaneer fleet with English captains from Jamaica and Barbados. Among them were John Morris, David Martien, Edward Mansfield, Robert Searle, and Henry Morgan. However, in 1663, Myngs was recalled to England to command a fleet in the war against the Dutch Republic and Henry Morgan assumed the role as the leader of all buccaneer activity in Port Royal.

1663–1670

Henry Morgan was king of the buccaneers

Unquestionably the most famous privateer of this era, Henry Morgan was the primary leader of the buccaneer fleets sailing for the English. He came to Jamaica in 1655 as a junior officer with the invasion force and quickly rose through the ranks to become a Colonel in the militia and Admiral of the Seas. Morgan was not the typical buccaneer captain. Always sailing with letters of marque, he was a close friend of the governor of Jamaica and owned three sugar plantations on the island. His uncle, Colonel Edward Morgan, was the lieutenant governor of Jamaica and Henry married his daughter, who was Henry's first cousin. Operating as captain from 1663 to 1670, Morgan's many raids included Portobelo in 1668, Maracaibo in 1669, and Panama in 1670. Among Morgan's many captains was Robert Searle, who would be the first pirate to attack Florida.

Chapter 8
Captain Robert Searle, Florida's First Pirate

The year was 1668 and the Spanish city of St. Augustine had remained unmolested for 82 years. The first attack on the city was led by Francis Drake who successfully attacked St. Augustine in 1586. Drake was operating as a privateer in support of England's continued military and naval operations against the Spanish. However, the second attack of St. Augustine was led by a pirate. His name was Robert Searle. Some accounts say that Searle decided to attack St. Augustine in the spring of 1668 in retaliation for the Spanish attack on the English colony on New Providence, in the Bahamas. This is doubtful because the Spanish attack on the colony occurred in 1664, four years earlier. A much more believable reason was greed. But who was Robert Searle and what became of him? More importantly, as the first pirate to attack Florida, what were the circumstances surrounding that famous attack?

1642–1651

The English Civil War resulted in the execution of King Charles I in 1649 and temporarily ended the monarchy. England was governed by the Commonwealth under the personal rule of Oliver Cromwell as Lord Protectorate from 1653–1658.

So far, no record of Robert Searle's early life has been found, but it is very likely that he was either the brother or a close relative of Daniel Searle, who was the governor of Barbados in the 1650s. During that time, Thomas Modyford also lived on Barbados and was a bitter political rival of Daniel Searle. This was the decade after the English Civil War and the nation was very politically divided between those who supported the royal family living in exile and those who supported the Parliamentary government under the rule of Oliver Cromwell. This political divide existed in the English colonies as well as in England. Modyford secretly supported the king while Searle backed Cromwell and the Parliamentary government called the Commonwealth.

The restoration of King Charles II in 1660 obviously caused major political turmoil throughout England and the English colonies. Many political leaders who had been supporters of the Parliamentary government were replaced with those who had remained loyal to the king. But over on Jamaica, things seemed to be going fairly smoothly. The buccaneers of Christopher Myngs were too busy taking Spanish cities to worry about politics. One of their most profitable raids was the 1662 amphibious assault on the city of Santiago, Cuba. But far more significant to our story is that the detailed records made of that raid mentioned the names of all of Myngs' captains and Robert Searle enters the official record for the first time.

Searle must have been with Myngs for some time before that because he was given command of an 8-gun sloop, which Searle named the *Cagway*. Searle obviously named his sloop after the original name of Port Royal which was called Cagway during parliamentary rule. That may have been a subtle way of showing his support of the defunct government. Another captain on that raid was Henry Morgan, who knew Robert Searle well. Shortly afterwards, Myngs was recalled to England and Morgan took over as the buccaneer leader.

Meanwhile, in 1664 King Charles II of England was attempting to patch up relations with Spain and decided to discourage buccaneer raids against them in the Caribbean. He also realized that he needed to change the leadership of Jamaica to a governor who would support his orders. Thomas Modyford, who had remained loyal to the royal family, was brought over from Barbados and appointed governor of Jamaica. Shortly afterwards, King Charles II sent a letter to Modyford ordering him to end privateer attacks against the Spanish.

1664–1671

Sir Thomas Modyford was governor of Jamaica

Robert Searle always seemed to be in trouble with Governor Modyford. This is easily explained if Robert was indeed a relative of Daniel Searle, Modyford's old political foe from Barbados. In the summer of 1664, Searle became the first privateer to be officially reprimanded by the new governor for violating the recent policy of King Charles II. In the summer of that year, Searle was at sea, operating as a privateer with letters of marque from the old Governor of Jamaica. He took two very rich Spanish prizes off Cuba and brought them to Port Royal. He planned to sell the cargo, divide the profits with his crew, and pay the English government their share in accordance with his commission. Most of the cargo had already been sold when King Charles II's letter ending attacks against the Spanish arrived at Jamaica. As Governor Modyford read the King's letter, he could glance toward the dock and see the sight of the two Spanish vessels being looted by Searle and his crew. It was instantly clear to Modyford that this would be the first test case and Robert Searle had to be publicly dealt with. An emergency meeting of the Council of Jamaica was called and the situation with Searle was debated.

The decision of the Council was that the Governor of Cuba should be informed and that the Spanish vessels and all the treasure and cargo should be immediately returned to the Spanish authorities. They also resolved to declare anyone a pirate who took Spanish prizes in the future and to take Captain Searle's commission from him. Additionally, they confiscated his rudder (navigational charts) and all his sails. In one of his letters, Modyford wrote, "On reading the King's letter of June 15 last, commanding restitution of captured ships and goods to the Spaniards: ordered that the ship and bark [barque] brought in by Captain Searle of the Port Royal be seized and restored to that nation, and also all specie [spice] that can be found; that notice thereof be sent to the Governor of Havana; that persons making any further attempts of violence and depredation upon the Spaniard be looked upon as pirates and rebels; and that Captain Searle's commission be taken from him, and his rudder and sails taken ashore for security." After that incident, Robert Searle often went by the name of John Davis. Perhaps he used that alias to hide his identity while he continued taking Spanish ships.

Searle's status changed rapidly in 1666 as the war between England and the Dutch Republic intensified. Governor Modyford needed as many privateers as he could get. Additionally, tension with the Spanish was increasing, so issuing letters of marque against the Spanish was now allowed. Robert Searle's ship, rudder (navigational charts) and sails were returned to him, and he joined an expedition under the command of Colonel Edward Morgan, the new Lieutenant Governor of Jamaica and the uncle and Father-in-law of Henry Morgan (Morgan had married his first cousin). With a force of nine vessels and 650 men, they sailed from Port Royal in March or April of 1666 to attack the Dutch islands of St. Eustatius and Saba. Unfortunately, during the attack, Edward Morgan died of heat stroke and the entire command split up. Searle formed a partnership with Captain Stedman, one of the other captains, and they sailed away. A year later, in the spring of 1667, Searle and Stedman in their two vessels with a force of about 80 men attacked the Dutch island close to Trinidad. They were very successful and completely sacked the entire island.

In May of 1668, Searle was on his ship, the *Cagway*, cruising off the coast of Cuba near the Florida Straits with letters of marque from Governor Modyford allowing him to attack Spanish shipping. He had another vessel with him, perhaps it was his partner Steadman, but this is only speculation. The two privateers spotted two Spanish vessels, a ship bound for Vera Cruz and a brigantine bound for Havana. The Spanish vessels were easily taken and Searle decided to keep them to add to his fleet. Aboard the Spanish brigantine, Searle questioned a passenger who was a former French surgeon named Pedro Piques. He told Searle about a vessel carrying a shipment of silver bars that had recently sunk near St. Augustine. He continued his tale by adding that all the bars had been recovered and that they were being kept in the treasury of the fort. This sounded almost too good to be true, so with his fleet of 4 vessels, Searle sailed north to the only large city in Florida.

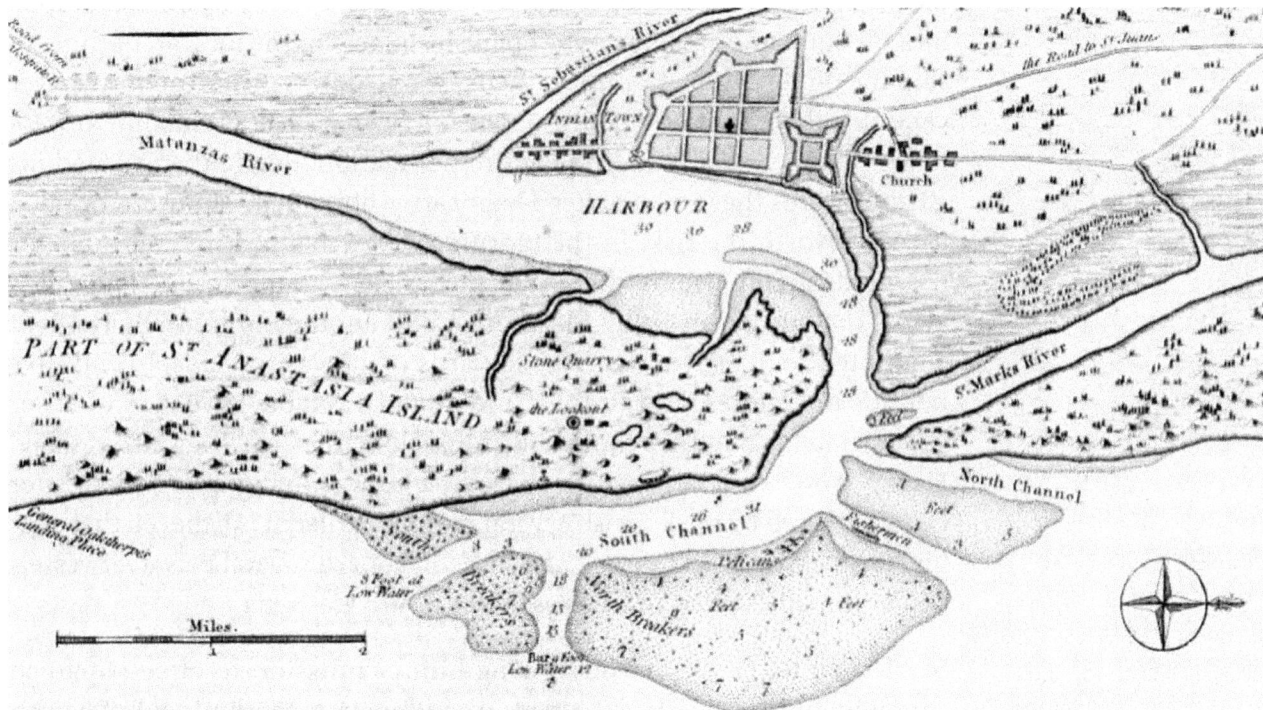

Figure 7: *St. Augustine 18th Century Map*

On the morning of May 28, 1668, Searle's fleet arrived at St. Augustine. He kept his two well-armed and English-built vessels out of sight, as they would be easily recognized as privateers as soon as they neared the town. Then, he sent a portion of his pirates towards the harbor aboard the two recently captured Spanish vessels. Searle believed that the town officials would not be alarmed at the sight of two innocent looking Spanish merchant vessels coming into port. The plan was working and the Spanish soldiers manning the guns of the fort were not concerned as Spanish merchant vessels entered the port all the time. The two vessels dropped anchor and waited for night. The harbor pilot rowed out to greet them and was told by one of the Spanish-speaking buccaneers that they were merchants from Mexico. The pilot returned to shore, reassured that they were legitimate Spanish merchants. No one in the town or at the fort was suspicious. Just after midnight on May 29, 1668, the other two vessels silently sailed past the fort's guns and quietly slipped into the harbor. However, they were spotted by Corporal Miguel de Monzón, who happened to be out in the harbor fishing. Realizing they were English privateers, Monzón frantically rowed ashore under fire from Searle's men. Wounded twice, he managed to make it to shore and alerted the Spanish authorities. But he was too late. As he shouted the alarm, Searle's buccaneers swarmed ashore and ran through the streets looting everything.

Anyone who resisted was shot. Some made it to the fort for protection including Governor Francisco de la Guerra de la Vega while others escaped into the surrounding forest. The Spanish soldiers in the fort did nothing to stop the pirates. In fact, some of them even fled into the forest. It was over very quickly. About 60 Spanish townsfolk were killed and about 70 were taken prisoner, nearly half the entire population. Searle then turned his attack towards the fort, but after a halfhearted attempt to capture it, decided it wasn't worth the risk. In the morning, he exchanged his prisoners for firewood and provisions, then he burned the city to the ground and sailed away. The buccaneers only lost 11 men and another 19 injured. The loot amounted to 133 silver marks, 760 yards of canvas for sails, twenty-five pounds of candles, and untold amounts of valuable jewelry and other personal items that they took while ransacking the city. They also rescued Henry Woodward, an English surgeon who was living in the town after being captured by the Spanish several months earlier. It is believed that after his rescue, Woodward served as a surgeon aboard several privateer vessels and eventually settled in Charles Town, South Carolina.

Searle had letters of marque giving him permission to attack Spanish shipping, but he clearly exceeded his authority when he took a Spanish port. That violation changed his status from privateer to pirate. Fearing that Governor Modyford would again be upset with his actions, Searle decided to quietly return to Jamaica and put in at Port Morant, which was just outside of the Governor's legal jurisdiction, until he could be sure of the governor's disposition. It was a good call, because Governor Modyford became furious when he learned of Searle's sacking of St. Augustine. Eventually Governor Modyford was informed of his arrival and on March 18, 1670, he wrote, "There arrived also at Port Morant the *Cagway*, Captain Searle, with 70 stout men, who hearing that I was much incensed against him for that action of St. Augustine, went to Macarry Bay and there rides out of [my] command. I will use the best ways to apprehend him, without driving his men to despair."

When Searle finally came ashore within jurisdiction of the governor, he was arrested. Governor Modyford wrote the English authorities stating that Searle was still in the custody of Jamaica's Provost Marshal awaiting trial and asked them what they wanted done with Searle and his men. After several months without answer, Searle was released and his vessel, the *Cagway*, was returned to him. It seems that the governor needed Searle and his men to assist in one of the greatest privateer ventures of all time. Henry Morgan was planning to take the immensely wealthy City of Panama.

The mark is not a coin but an accounting value. One mark is equal to two-thirds of one pound in weight. 133 silver marks equals approximately 89 pounds of silver.

It was the summer of 1670 and the Treaty of Madrid had just been signed which ended hostilities between England and Spain. Even though the news of the Treaty reached Port Royal before Morgan left, Governor Modyford and Morgan didn't take the news seriously and decided to go ahead with their plan to take the City of Panama anyway. Morgan obtained his letters of marque from Modyford and left Port Royal accompanied by his old comrade, Robert Searle. Morgan had assembled the largest buccaneer fleet in history. He began his careful amphibious operation landing on the Isthmus of Panama on December 27, 1670. Moving up the Chagres River and crossing to the Pacific side, Morgan led a well-organized attack of six ranks of infantry with both flanks supported by 200 cavalry. As Morgan's main force of buccaneers opened fire, the Spanish suffered heavy casualties and the rest of the troops routed. Panama City fell, and Morgan had taken an estimated 400,000 pieces of eight.

DECEMBER 27, 1670

Henry Morgan takes Panama City

After Panama was captured, Morgan gave the vitally important task of securing the port to his old comrade, Robert Searle. He was ordered to capture any vessel attempting to leave the port. Upon reaching the shore, Searle's men found a Spanish barque, the *Fasca*, which had been intentionally grounded and set on fire during the attack to prevent the English from capturing it. But Searle's men managed to put out the fire and refloat the vessel. In a few days, Searle managed to capture three other Spanish vessels. With his small flotilla, Searle cruised the offshore islands of Perico, Taboga, Tobogilla, and Otoque where they took many prisoners and a lot of valuable property. Unfortunately, that included a large amount of Peruvian wine and by evening all Searle's men were drunk, too drunk to notice the large Spanish galleon approaching the port. It was the *Santissima Trinidad*, heavily laden with gold, silver, pearls, jewels, and other most valuable goods. The galleon had been in port when Morgan attacked Panama and was ordered to flee to another port, but the captain decided to simply put to sea and return to Panama after he felt the English privateers had left.

The captain of the Spanish galleon, Don Francisco de Peralta, saw Searle's barque but assumed it was just another Spanish vessel that had escaped the attack of the city. The galleon anchored nearby and a small party of seven men was sent ashore in a rowboat to get fresh water. Some of Searle's men accidentally ran into the Spanish shore party,

Figure 8: *Morgan Attacks Panama*

captured them, and brought them to Searle. Under threat of torture, the prisoners told Searle of the nearby location of their galleon. Additionally, they told Searle that the galleon was armed with only seven guns and about a dozen muskets.

Realizing that this incredibly rich prize was theirs for the taking, Searle quickly ordered his men to set sail, but they were all still too drunk to respond, so Searle decided to wait for morning. However, during the night, De Peralta became suspicious when his men didn't return, so he set sail and escaped in the darkness. When Morgan learned that Searle had let this rich prize slip through his fingers, he was outraged. Searle was bitterly reprimanded, and Morgan never trusted him again. After returning from Panama, Searle left Morgan's company and began operating in the Gulf of Campeche. Several years later, according to William Dampier, he was killed in a duel with one of his men on a tiny island at the northern end of the Gulf, still known as Searle's Key.

Chapter 9
Buccaneers Become Pirates

The Treaty of Madrid in 1670 put an end to large-scale state sponsored piracy in the Caribbean and the Pacific. It also put an end to buccaneers being well-respected citizens. The buccaneers would have to become true pirates, at least for a little while. This is the second era of the "Golden Age of Piracy," the buccaneer pirates. It was a time when privateer commissions issued through letters of marque soon became very restrictive and far less profitable. Many buccaneer privateers simply turned to piracy and became buccaneer pirates while others became pirate hunters.

The constant wars in Europe added to the confusion; your friend on one venture is your enemy on the next. A stronger royal naval presence in the Caribbean made things far less profitable for those who chose to remain operating legally as privateers, as they would now have to share a substantial portion of the prize with the naval commanders. Also, they had to follow orders from royal naval authorities. The amphibious assaults that the buccaneers had been so famous for were now exceedingly rare. And for the buccaneers who chose to become pirates, the stepped-up naval protection for merchant fleets made prizes harder to come by and even harder to take.

The huge buccaneer port of Tortuga faded into obscurity and Port Royal turned hostile towards the buccaneer pirates. However, support for pirates still existed among the citizens and merchants all along the Atlantic coast of North America until the end of the 17th century. For local merchants, the need for goods and money far outweighed the way in which they obtained them. Corrupt government officials cavorted freely with pirates and sponsored their ventures. Pirates openly traded in the streets of ports like Charles Town, Marcus Hook, Newport, and New York.

Marcus Hook is located about 20 miles down the Delaware River from Philadelphia

But profits for those pirates were down when compared to the earlier days. More and more, the buccaneers began to leave the troubled waters of the Caribbean. Many pirates tried their hand at taking prizes along the North

1683-1691

William Dampier's round the
world voyage

American coast while others sailed the waters of the Pacific. Since sailing around Cape Horn was exceptionally dangerous, most pirates crossed the Isthmus of Panama over land, stole some vessels on the Pacific side, and began operating along the Pacific coast.

Among the most famous of these pirates was William Dampier, who was a crewman with several different pirate captains including Edward Davis in the Pacific in the 1680s. Eventually sailing across the Pacific, Dampier felt the time was right to leave these pirates and quietly slipped away near Nicobar Island just west of Thailand. Finding passage on an English merchant vessel, Dampier returned to England in 1691, and wrote a book about his adventures titled "*A New Voyage Round the World.*" Today, it is still in print but entitled "*Memoirs of a Buccaneer.*"

1693

English pirates begin raiding
in the Indian Ocean

However, pirate ships sailing across the Pacific or into the waters of the Indian Ocean were rare. Those oceans were too far away for pirates to reach safely. This rapidly changed in 1693 when Adam Baldrige established a pirate friendly port on the island of St. Mary's off the coast of Madagascar. Pirates could stop there to resupply and repair their vessels in safety on the way to the Indian Ocean and on the return voyage. For a short time, the waters of Madagascar became the center of pirate activity. Among the most famous of these Indian Ocean pirates operating in the mid-1690s were Thomas Tew, Henry Every, and William Kidd. Tew was killed in 1695 while attacking a vessel. Every is believed to have retired with a fortune in stolen loot. And Captain William Kidd, the most famous of the pirates from the Indian Ocean, was executed in 1701.

For English pirates, the waters of the Indian Ocean rapidly faded as a lucrative place to operate after the pirate port of St. Mary's had begun to decline. Its founder, Adam Baldrige, was forced to flee when he got into a dispute with the local natives. Finally, in 1699, warships from the English Royal Navy attacked and destroyed the pirate port which closed it for good.

Chapter 10
Pirate Raids of the 1680s

In the late 16th century, Spain divided Florida into four provinces and began to establish dozens of Franciscan Missions throughout each of them. Those provinces were named for the predominant Native American tribe living within each province. The Apalachee Province was comprised of what is now the panhandle of Florida. The Timucua Province extended across the middle of the state, running from the St. Johns River west to the Suwanee River. On the northeast coast, the Mocama Province began at the mouth of the St. Johns River at present-day Jacksonville and extended north to St. Simons Island in Georgia. The Guale Province was located entirely in present-day Georgia and ranged from Cumberland Island to St. Catherine's Island, located about 20 miles south of the city of Savannah. Each province contained between 15 and 40 individual missions and thousands of tribal villages.

The English colony of Carolina was established in 1670 when the city of Charles Town was founded on the west bank of the Ashley River. As the population grew, the city was relocated to its present location a few miles to the southwest and across the river.

Serious problems began for the Spanish missions and for the people of St. Augustine in 1670 when the English established the colony of Carolina with the city of Charles Town as its capital. The Spanish didn't like it, but there was nothing they could do to drive the English out. For the main government back in Spain, there were too many other problems that took priority.

Figure 9: *San Marcos, St. Augustine*

Manuel de Çendoya, the governor of Florida in St. Augustine, would just have to deal with the situation on his own. In 1671, Çendoya sent 25 soldiers to protect the Mission Santa Catalina on St. Catherine's Island. That was their closest mission to Charles Town. But a far more critical concern in the 1670s was attacks from the Westos Indians, who constantly raided both English plantations and Spanish missions. To defend the largest Spanish city in Florida, construction of a highly imposing fort began in 1672. Named the Castillo de San Marcos, it took 23 years to complete. Today the fort still exists in excellent condition as one of the nation's National Parks.

Charles Town desperately needed supplies and other goods that they couldn't get from England, goods like cloth and tools. England severely limited trade with the colonies and Carolina was no exception. The answer for the fledgling colony was pirates. With total disregard for English trade laws and tariffs, pirates could bring in all sorts of commodities to any settlement or port whose residents didn't mind dealing with them. Charles Town quickly became one of those ports. As pirate friendly ports in the Caribbean were rapidly disappearing, Charles Town emerged as the best place to sell their stolen goods. By the early 1680s, the Carolina port of Charles Town had become a pirate haven. But these pirates were attracted to Charles Town for other reasons than just being welcome. Nearby Florida was developing as an ideal place to conduct raids and to attack shipping. As St. Augustine grew, the number of ships carrying valuable cargo increased along the Florida east coast.

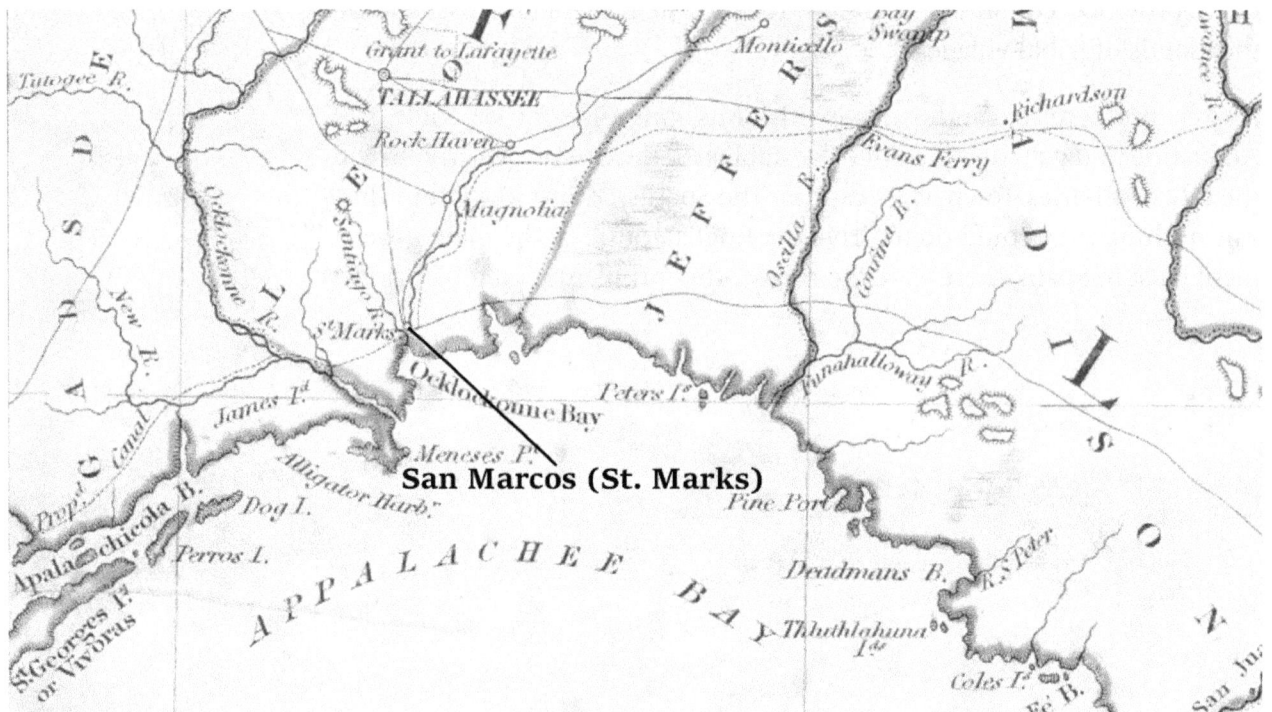

Figure 10: *Apalachee Bay*

The Franciscan missions along the east coast of Florida didn't have very many gold or silver valuables, but they did have substantial amounts of food. Each mission had several large ranches associated with them as well as extensive farms. A portion of the commodities was distributed among the natives, but the rest were sent to other Spanish colonies that were in great need of food. In many of the Caribbean colonies, farmland was primarily used to grow valuable cash crops such as sugarcane or tobacco. Those colonies had to rely on other colonies to produce crops that they could eat such as grain, rice, and vegetables. Florida was perfectly situated to supply them and became one of the primary suppliers of food for the Spanish in the Caribbean.

The Real Situado was a Spanish term that referred to a year's amount of food, supplies, and money intended to feed, equip, and pay the salaries of a detachment of soldiers.

Beginning around 1680, pirates began raiding Spanish missions for the food they had. Not just to eat themselves, but to sell in Charles Town. This was encouraged by the residents of the English colony for two reasons. First of all, they needed the food. But more importantly, they intended on eventually extending their territory southward into Florida and weakening the Spanish missions seemed like a good start. These small attacks were devastating to the local Spanish economy. But of far greater impact was the loss of their supply ship which was carrying the city's Real Situado. Once a year, Spain sent a ship carrying food and supplies for the fort as well as all the money needed to pay the salaries of the soldiers. This shipment was called the Real Situado and was large enough to cover expenses and needs for an entire year. In 1682, the supply ship was captured by English privateers and everything was taken. But pirate raids on Florida weren't limited to just the east coast.

Jamaica was in need of slaves for its plantations. In the 1680s, English pirates raided Apalachee villages along Florida's Apalachee Bay on the northwest coast looking to capture some of the natives and turn them into slaves. Spanish missions in the region also attracted pirates in search of food. The Spanish had established a mission on the St. Marks River back in 1639. Aware of potential pirate raids, the mission was reinforced by a wooden stockade built at the confluence of the St. Marks River and the Wakulla River. Even though it was constructed of wood, the walls were painted to look like stone in an attempt to make the tiny fortification appear stronger than it was.

In the early 1680s, the French were also operating in the area looking to establish a colony near the mouth of the Mississippi River. A band of French privateers located the fort of San Marcos and decided to approach. At first, the fake stone walls fooled them and they debated over making an attack, but eventually they realized the deception and quickly took the fort. They held the friars for ransom and a message was sent to the main Spanish mission in the province, Mission San Luis de Apalachee in present-day Tallahassee. But Franciscan friars are vowed to poverty and the mission had no money. When the reply came explaining the situation, the French politely released the hostages and left. The English weren't so kind. A year earlier in 1681, English privateers raided the nearby ranch of La Chua. They returned in 1682 and conducted a second raid. Then, after the French left the fort of San Marcos de Apalachee intact, the English privateers attacked it and destroyed the fort completely.

After 15 years, a band of English pirates once again decided to attempt to attack St. Augustine itself, just like Robert Searle had done. On the morning of April 5, 1683, several small boats loaded with pirates quietly approach the port. Spanish soldiers were far more observant than in the past and the boats were spotted. But the pirates also spotted the Spanish and decided that their plan wouldn't work. They rowed back to their ships and sailed north. Not wanting to leave Florida waters empty-handed, they attacked several Spanish missions including San Juan del Puerto near present-day Jacksonville and San Felipe on Cumberland Island.

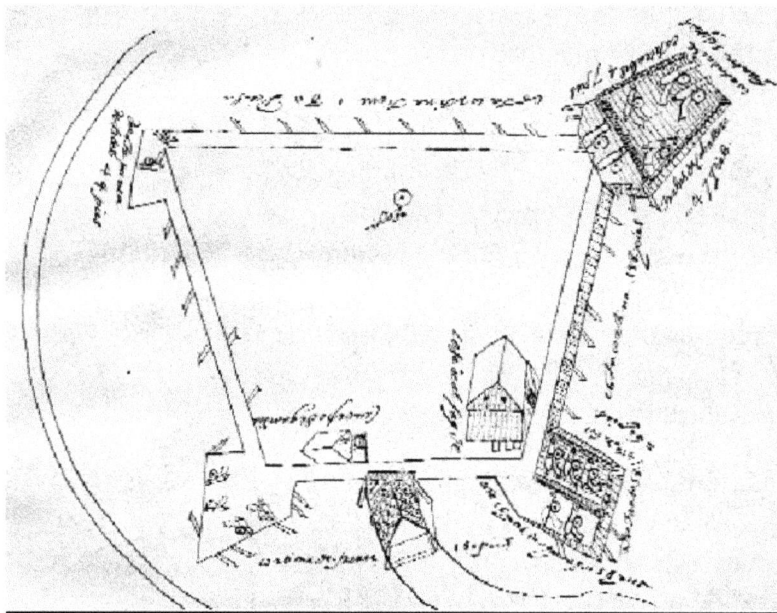

Figure 11: *San Marcos de Apalachee*

English pirate attacks on missions in the Guale Provence intensified and by 1684, the province was nearly destroyed. In August that same year Governor Marquez Cabrera of Florida managed to persuade most of the Guale and Mocama Indians to relocate to Mission San Pablo south of the St. Johns River for protection. In October of 1684, English privateers raided Guale again and destroyed Mission San Juan del Puerto near the mouth of the St. Johns River and Mission Talapo on Sapelo Island. They also raided other missions and villages throughout Guale Province. Spanish soldiers did not engage the larger force, and the entire population retreated inland, leaving the villages to be plundered and destroyed. The local inhabitants decided to follow the governor's guidance and went south. By the spring of 1685, Guale as a Spanish province disappeared.

During one of those raids, one of the English pirates was captured. Upon questioning, he revealed that there was a pirate fleet at Charles Town that was comprised of a combination of English, French, and Dutch buccaneers. This fleet was under the command of Michel Sieur de Grammont who was among the most successful buccaneers in the Caribbean. A French privateer operating out of San Domingue, Grammont was of noble blood. It is believed that he turned to piracy after a duel forced him to leave France. During his 15-year career, he led successful raids on the Spanish ports of Maracaibo, Gibraltar, Trujillo, La Guaira, Cumana, Puerto Cabello, and Veracruz. His partner in this venture was another French buccaneer named Nicolas Brigaut.

In St. Augustine, the rumors of the impending attack were confirmed in February 1686 when a Dutch merchant from New York arrived in port. He knew about the fleet at Charles Town and told the Spanish authorities. The city went on alert. But the pirates never arrived. In April 1686, while sailing towards St. Augustine, Grammont's fleet ran into a storm. Brigaut's ship was wrecked on Matanzas Inlet where he and his survivors were found by Spanish soldiers and executed. Grammont's ship is presumed to have sunk at sea, as he was never seen again.

In retaliation for all the English attacks, Spanish authorities in Cuba authorized Spanish privateer Alejandro Thomas de Leon to attack Charles Town. In May 1686, de Leon arrived at St. Augustine with two ships then sailed towards the Carolina coast. He decided not to attack Charles Town, as it was too well defended, so he chose to attack another nearby settlement located at present-day Port Royal, South Carolina. However, the settlers were warned and fled to Charles Town. The Spanish burned the settlement to the ground then raided and destroyed several nearby plantations. During those raids, the Spanish recovered some religious artifacts previously taken by the English. They also took all the possessions of the plantation owners. On their way back to St. Augustine, the two vessels ran into a storm. De Leon's ship sank, and he drowned while the other ship managed to make it to port.

Other English raids in Florida continued through 1686, but those attacks began to create a political problem for the colony's governor, Joseph Morton. Most of the colony's citizens were in favor of the attacks and wanted them to continue, believing that they might eventually take control of more land to the south. But the governing Grand Council and the colony's Proprietors, who actually ruled the colony, feared Spanish reprisals and wanted peace. In July 1687, the governor was replaced by James Colleton who sided with the Proprietors and all pirate raids quickly ended. All hostility on both sides temporarily ended in 1688 when England and Spain became allies in the War of the Grand Alliance against France.

Chapter 11
Ranson's Miracle

One of the more interesting stories associated with the long and colorful history of St. Augustine is the story of the English pirate, Andrew Ranson. He was captured in 1684 along with 10 others during one of the raids on a Spanish mission. Ranson was born in Newcastle, England around 1650 and came to the West Indies sometime in the 1670s. A professional sailor and possibly even a ship's pilot, he married and settled with his wife on New Providence, Bahamas sometime prior to 1683. In reprisal to a 1664 Spanish attack, the Governor of the Bahamas issued letters of marque to Captain Thomas Jingle to attack Spanish shipping and ports. Jingle had 5 vessels in his fleet and Andrew Ranson signed aboard one of them. Just before sailing, they were joined by another English vessel from Charles Town that had participated in the failed attack of St. Augustine the previous year.

Shortly after the six vessels left New Providence, they captured a Spanish frigate, the *Plantanera,* that was on its way from St. Augustine to Vera Cruz to collect the Real Situado for the fort of San Marcos. Near Biscayne Bay, which is located near present-day Miami, they also captured a Spanish scout vessel that was gathering intelligence on English pirates for the governor of Cuba. The crew members of the scout vessel were mildly beaten to gather information about the nearby Spanish settlements. During their questioning, another vessel piloted by Ranson came alongside. Ranson boarded the main vessel and joined in the questioning by hitting one of the captured crewmen with a stick. That crewmember was a Spanish sailor named Miguel Ramon. All the Spanish captives were eventually released.

A short time later, they were joined by five other privateer vessels. The captains met and decided they had a strong enough force to attack St. Augustine. But just as the privateer vessels approached the prize city, a severe storm scattered them in several directions. Five of the vessels managed

A ship's pilot was responsible for navigation and was a highly respected leadership position onboard all vessels.

to regroup near the St. Johns River and decided to give up their plan to attack the city. However, they were in great need of food. Ten men under Ranson's command were sent ashore in a small pirogue to look for corn or anything else they could eat.

A pirogue is a flat-bottomed canoe ideal for navigating backwater marshes and bayous.

Unfortunately for Ranson and his men, the Spanish were alerted and 50 soldiers were quickly dispatched from St. Augustine to intervene. Ranson and his men were easily captured and then taken to the St. Johns River where the Spanish soldiers forced them to signal to their vessels anchored offshore. The Spanish soldiers stayed hidden nearby and hoped to lure some of the Englishmen ashore so they could be captured too. The plan seemed to be working as a longboat filled with privateers soon approach the shore. But at the last minute, the privateers suspected an ambush and swiftly returned to their main vessel. Ranson and his men were taken back to St. Augustine to stand trial.

At his trial, Ranson insisted that he was just one of the crew but some of his men identified him as their leader in return for a lighter sentence. Then, the Spanish prosecutor produced Miguel Ramon, the sailor whom Ranson had beaten earlier. He identified Ranson as the ship's pilot and one of the men who had beaten him. As a pilot, Ranson would most certainly be the leader of the group. Ranson's men were convicted and sentenced to hard labor for life and Ranson, as their leader, was sentenced to death. Negotiations between England and Spain eventually led to the release of the others, but Ranson was still to die. While imprisoned in San Marcos awaiting execution, Ranson began speaking with the Franciscan priest named Perez de la Mota, and converted to Catholicism.

The usual form of execution used by the 17th century Spanish was the garrote. In this method of execution, the prisoner stands with his back to an upright board and a rope is placed around the prisoner's neck. The rope is tightened behind the board and strangles the prisoner to death. It normally takes about 6 twists to complete the process. In Ranson's case, as the rope was tightened and Ranson lost consciousness, the rope suddenly broke. Ranson quickly regained consciousness and the rope was examined. It was a new rope with no signs of wear. The only explanation for the rope breaking was offered by the priest, Perez de la Mota, who proclaimed it a divinely inspired miracle.

Over the next three years, while Ranson sat in prison, the governor and the priest debated over the circumstances of the execution. Father Perez de la Mota insisted that it was a miracle and demanded Ranson's release, while the governor believed it was just unusual circumstances and wanted him executed. This small affair quickly developed into a major rift between the church and the state that involved the Bishop in Cuba and even the Royal Council back in Spain. In 1687, Quiroga y Losada was appointed governor of Florida and he wanted to end the difficulty between the church and his office. While the matter was being settled, he granted Ranson parole if he agreed to work as a carpenter on the construction of the new fort of San Marcos.

By 1690, the case was finally brought to the authorities in Spain. On the king's orders, Ranson was released and Father Perez de la Mota was summoned to Spain to give testimony. The case was finally heard in 1692 when la Mota produced the actual rope and argued the miracle defense on Ranson's behalf. But Spanish bureaucracy was exceptionally slow and inefficient. No decision was made and eventually Ranson was returned to prison to await a verdict that never came.

Ranson's luck finally changed in 1702. Spain and England were at war again and the English from Charles Town laid siege to St. Augustine for two months. The new Governor, Joseph de Zuniga y Cerda, offered Ranson his freedom if he acted as an interpreter. When the siege lifted, Ranson was released and finally left St. Augustine with the English attackers.

In recent years, the validity of this story has come into question. Some pirate enthusiasts have even suggested that the entire story was invented in the late 20th century and that no historical documentation supporting this story exists. Those enthusiasts are incorrect. This story first appeared in 1960 as an article in *The Florida Historical Quarterly* titled *"Andrew Ranson: Seventeenth Century Pirate?"* The article was written by Leitch J. Wright and was very well researched. Some of the original source documents from the 17th and 18th centuries that are cited in the article include correspondences from Friar Perez de la Mota, Governor Diego de Quiroga y Losada, and Governor Jose de Zuniga y Cerda. I believe this story to be completely true.

Chapter 12
Did Captain Kidd Visit Florida?

Florida has several cities that claim to have a long tradition of pirates using their ports as a base or of them burying their treasure near the beaches. Tourists who visit these cities will generally be told that Captain Kidd buried some of his treasure nearby. Most notably is the city of Amelia Island, located in the northeast corner of Florida, along the Georgia-Florida border. For many years, stories of Kidd's connections to this island have been told by locals and have even been included in some of the tourism websites and in tourist brochures. But did this famous pirate actually stop at Amelia Island or anywhere along the Florida coast and bury treasure?

There is no doubt that Captain William Kidd is among the most famous pirates of all time. But just like so many other pirates of the period, very little is known about his early life. Some accounts are based upon incorrect or even fictitious information. The most recent scholarly research states that William Kidd was born in 1654 in Scotland, perhaps near Dundee. He settled in New York sometime before 1688. The first actual record of Kidd lists him as a member of a privateer crew sailing the Caribbean in 1689. During that cruise, the record tells us that Kidd was made captain of a captured French ship which he renamed the *Blessed William*. Kidd operated primarily in the waters around Nevis and didn't sail on his own. During that war, privateers were kept under far more control than in previous wars. Kidd sailed alongside of a Royal Navy Squadron and followed the orders of the squadron commander. After his service was complete, Kidd was given the sloop *Antigua* to command and he returned to New York where he assisted the new governor, Colonel Henry Sloughter, in putting down a local rebellion.

Kidd rose to social prominence in New York City, where he married well, built a large house near modern day Wall Street, and made all sorts of

1696

Captain William Kidd and the *Adventure Galley*

political connections. In 1695 he began working on a plan with several prominent New York businessmen and the governor of Massachusetts to help him get a privateer commission directly from King William III. England and France were at war and a privateer commission would make any French vessel a legal prize. His plan was to look for rich French ships in the Indian Ocean. The king not only gave him the commission but gave him a brand-new warship, the *Adventure Galley,* and Captain Kidd set sail.

But Kidd seemed to have very bad luck. After looking for prizes in the Indian Ocean for nine months, Kidd only took one vessel, and that prize was very meager. Kidd's luck seemed to change on January 30, 1698, when his crew sighted the largest and richest treasure ship ever seen. It was the *Quedagh Merchant,* which was a 500-ton Armenian merchant ship laden with gold, jewels, silver, silks, sugar, and guns. Kidd's *Adventure Galley* overtook and seized the ship but there was a problem. The *Quedagh Merchant* wasn't a French ship, it belonged to one of England's allies, and the captain of the ship happened to be English. But after searching through the ship's documents, he found a French letter on board that gave the treasure ship permission to safely sail through French-controlled waters. It was like a free pass. Kidd rationalized that this was enough to justify taking the prize as a French vessel. To Kidd, it proved that they were collaborating with the French.

Kidd stopped at the pirate base of St. Mary's off the Madagascar coast to repair his vessels and prepare for the long journey back to New York. Normally this would have only taken a month or two, but it was about a year before Kidd finally sailed for home. What was he doing all that time? The terms of his commission required that he give a significant portion of whatever valuables he took to the English government. So, in order to keep a larger portion of the treasure for himself, it is believed that Kidd sailed to the China Sea and buried a part of the loot on a tiny uninhabited island before sailing home. Upon Kidd's return to St. Mary's, he realized that the *Adventure Galley* was too badly damaged to make the return voyage.

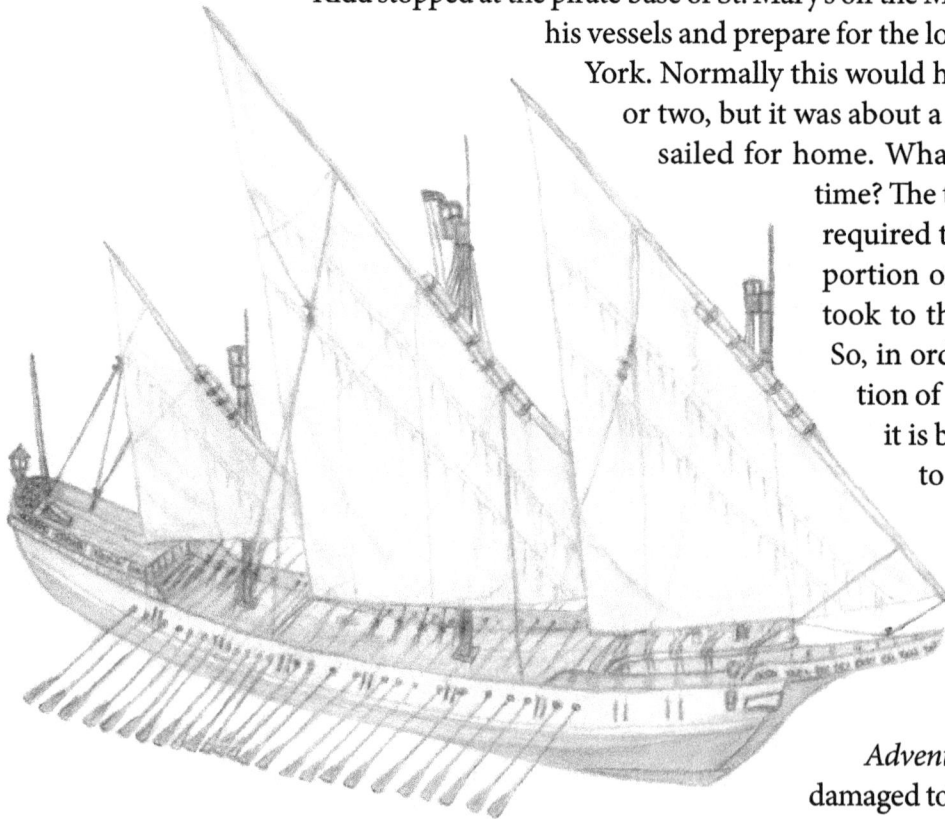

Figure 12: *The Adventure Galley*

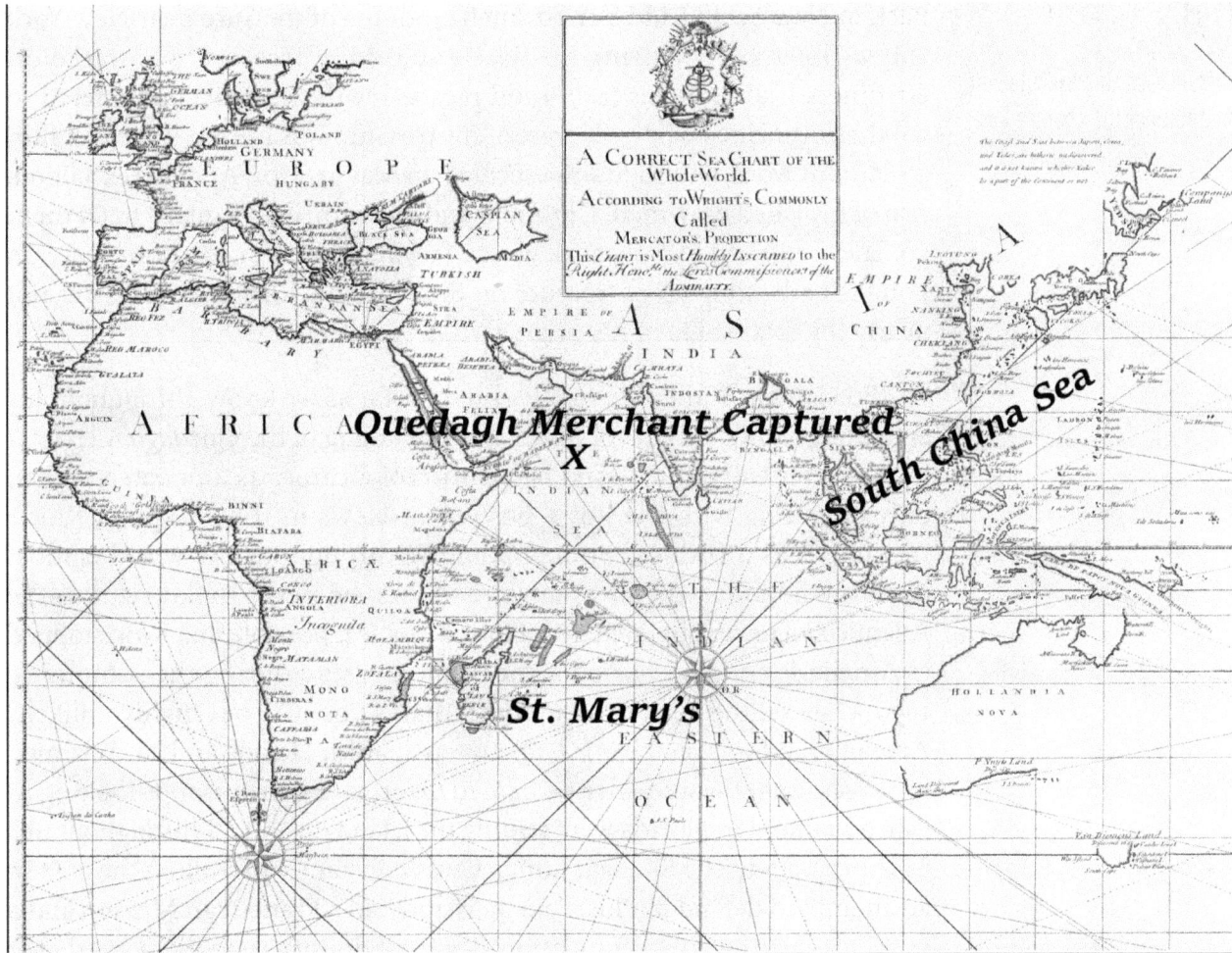

Figure 13: *Indian Ocean*

After abandoning the *Adventure Galley,* Kidd headed for home aboard the *Quedagh Merchant.* But to Kidd's surprise, when he stopped for supplies at Anguilla in April 1699, he learned that he was wanted for piracy. Kidd's taking of the *Quedagh Merchant* had created an international incident and was a great embarrassment to the King of England. Additionally, the war with France was over, so Kidd's commission was invalid.

Confident that he could clear his name in court, Kidd sailed the *Quedagh Merchant* to Catalina Island, a tiny island off the southeast coast of Hispaniola. He obtained a sloop named the *Antonio* from one of his old buccaneer friends and transferred most of the loot aboard. Leaving the *Quedagh Merchant* behind with a portion of the loot and a few of his crew to guard it, Kidd sailed back to New York onboard the sloop *Antonio.* Of course, as soon as Captain Kidd sailed out of view, the crewmen took all the loot and scuttled the ship. The wreck of the *Quedagh Merchant* was actually discovered 70 feet offshore of Catalina Island in 2007.

1701

Captain Kidd was executed on mud flats of the Thames River

Back in New York, Kidd buried small amounts of treasure near New York City as insurance. During his trial, £20,000 of treasure was found on Gardiners Island. At his trial, Kidd pleaded with the court and offered to tell the authorities where the rest of his treasure was buried if they let him go. It didn't work, and Kidd was executed on May 23, 1701. After his execution, two of his jailed crewmen, Churchill and Howe, bribed a guard to let them go. They returned to New York where they recovered £1,500 of treasure on Long Island and £700 of treasure on Block Island. This of course, was not nearly the total value of his original loot.

So, did Captain Kidd ever visit Florida? As far as we know, throughout his entire life, Kidd's only act of piracy was the taking of the *Quedagh Merchant* in 1698. He really didn't even have much of a career as a privateer either, only operating with the Royal Navy near Nevis in 1689. Captain Kidd's reputation as a ruthless pirate operating throughout the world is most certainly fictional. As far as anyone knows, he never took prizes in the Atlantic and never operated near Florida. But is it possible that Kidd stopped at Amelia Island and buried some of the treasure during the summer of 1699 while sailing back to New York? That is possible, but highly unlikely. Amelia Island was on the border between Spanish Florida and land claimed by England that would later be named Georgia. English troops looking to gain a foothold in Florida were raiding the island on and off throughout this time period and the Spanish military was constantly patrolling the area in an attempt to defend it. Plus, Amelia Island had a Franciscan Mission and a small Native American population. That just doesn't seem like a good spot to bury treasure, especially when there were lots of uninhabited islands all along the coast of the Carolinas that were far easier to get to. I would have to say that Captain Kidd did not visit Florida.

Chapter 13
Privateers of Queen Anne's War 1702-1713

At the end of the 17th century, everything changed for pirates and privateers. Public opinion swayed against them as the trial and execution of Captain Kidd sent strong repercussions through the continent. New laws were passed and government officials were replaced. By 1700, to be a pirate meant that you were a criminal operating under a death sentence. For privateers, opportunities to get letters of marque were severely limited. That rapidly changed in just two years.

In 1702, a major war broke out among just about all of the European nations. It was called the War of the Spanish Succession and was primarily over who would be the next King of Spain. When the Spanish King Charles II died with no children, the Spanish throne was claimed by two people, Archduke Charles, the son of Leopold I, ruler of the Holy Roman Empire, and Philip, Duke of Anjou, the grandson of King Louis XIV of France. Both were closely related to the former king. The war was fought on a very large scale with huge armies and naval engagements. Privateers also played a vital role. During the war, 128 privateer commissions were issued in Bristol alone.

For the pirates of the world, the opportunity to become legal privateers had returned. Immediately at the start of the war, pardons for all pirates were offered on both sides and letters of marque against their respective enemies were issued to anyone with a ship and a crew. This started the third era of the "Golden Age of Piracy," the privateers of Queen Anne's War. These privateers were generally all sailors, not a mix of sailors and soldiers as in the last century. Many captains and officers on merchant ships became privateers either out of patriotism or economic necessity.

But privateers were handled far differently than in the past centuries, they were far more controlled. Quite often, privateers were used to augment

1702–1713

War of the Spanish Succession

England, Dutch Republic, Holy Roman Empire against Spain, and France

1702

Queen Anne was crowned as Queen of England. On May 1, 1707, under the Queen's guidance, Parliament passed the Acts of Union, officially uniting England, Scotland, and Wales forming Great Britain. Consequently, after 1707, the English Royal Navy was referred to as the British Royal Navy.

Naval forces and sailed directly under their control. In other instances, reliable captains were selected to operate on their own but under very restrictive orders. Those privateers would operate much like royal naval forces, with strict discipline and a command structure similar to regular navies. Among the most famous and successful privateers during that time was Woodes Rogers, who sailed around the world between 1703 and 1707. The only privateers who were permitted to operate independently did so on a very small scale.

The anti-piracy trend that had begun in 1700 continued and intensified. By the war's end, local laws were toughened and piracy was no longer tolerated by local merchants in most of the major ports. Pirates were now looked upon as unpatriotic criminals. Privateers were still welcome in most of the ports, but they didn't receive the "hero like" fervor from the local merchants that they had enjoyed in the 17th century. Their celebrating and carousing were now viewed as poor behavior. New York and a few other ports welcomed privateers but restricted them to a few streets where their celebrating was controlled.

Right at the start of the war, an event took place that set the stage for the rise of pirates 13 years later. A major naval battle occurred near Vigo Bay, on the west coast of Spain just north of the Portugal border. The English Royal Navy soundly defeated the French fleet and Spain became concerned about an English attack of their treasure fleet. Spain decided to slow and eventually halt their yearly treasure fleet which sailed from Havana. But treasure still kept flowing into Havana from all of Spain's colonies in the New World and the Pacific. Even though Spain desperately needed money, the enormous wealth accumulating in Cuba would have to remain there until the war's end.

In America, the War of the Spanish Succession was called Queen Anne's War. Queen Anne was crowned in 1702, just as the war was getting underway. Queen Anne was well-liked by her subjects and was a particularly good ruler. Under her influence, the army and navy became modernized, making Britain among the most powerful nations of the 18th century. To the colonists in America, the war was far more obscure than in Europe and the meaning, purpose, and political goals were unclear. As far as the British colonists were concerned, Queen Anne had declared war on the French and Spanish once again and that was that. In 1702, the very first year of the war, a small Spanish invasion fleet attacked Charles Town, Carolina, and a few of the outlying settlements, but they were unable to do any significant damage. In retaliation, 500 militiamen from Carolina and 300 supportive Indians attacked the Spanish city of St. Augustine, Florida. After a two-month siege, they were unable to take the main fortress, but they did manage to burn the city. During that siege, Andrew Ranson, who had been held a prisoner by the Spanish for 18 years, was finally released and returned to Charles Town with the fleet.

In 1703, a combined force of French privateers and Spanish soldiers invaded the small English colony of New Providence in the Bahamas and totally overwhelmed the 250 English residents to capture the town. They returned to Cuba a few days later with 13 captured vessels and about 100 prisoners, including the island's governor, Ellis Lightwood. The Spanish and French again invaded New Providence in 1706. The main port of Nassau was sacked leaving only 27 families living in small huts. In response, English privateers began arriving in large numbers. The war was at its height and Port Royal, Jamaica, was too far away from the action, which seemed to be around Florida.

That same year, just as the English privateers were becoming organized at Nassau, a large, combined Spanish and French fleet sailed from Havana and attacked Charles Town again. But just as before, they were unsuccessful. The shots fired at the city's walls just seemed to bounce off. One interesting aspect of the city's defenses was that the protective walls were made of

Figure 14: *Nassau, New Providence 1793 Map*

Palmetto trees are native to the North American coastline along the Gulf Coast and up the Atlantic to North Carolina.

1707

English privateers develop Nassau as a base

1713

Treaty of Utrecht ends the war

King Philip V remains king of Spain

palmetto logs which are fairly soft and fibrous. They were ideal for use as defensive walls because the attackers' cannon shot could not break the wood apart. By 1707, Nassau emerged as a new privateer base. It was close to the main shipping lanes along the Florida Straits. Additionally, the waters around the town were too shallow for large warships and Spanish attacks would be limited to small vessels.

Small pirate or privateer attacks along the Florida east coast were rare during the war due to the large military presence from both the British and the Spanish. On the west coast, the few meager Spanish and French colonies along the Florida Gulf coast didn't attract shipping of any value and consequently privateer action in the Gulf of Mexico was almost non-existent. However, privateers must have had success taking shipping in the Florida Straits. It is possible that these Nassau privateers occasionally used some of the isolated inlets of the Florida east coast to hide and ambush passing prizes. But if they did, once a prize was taken, they would have returned to Nassau with their captured cargo and money.

The War of the Spanish Succession (Queen Anne's War) ended in 1713 with the signing of the Treaty of Utrecht. King Philip V kept the Spanish crown and most of the territory went back to pre-war status. However, the war had drained everyone's resources. At the same time, it had produced a large number of privateers who were now out of business. For the first time in 12 years, there was no war and consequently no letters of marque to be had. For many of them, privateering was the only life they knew. Many of them temporarily turned to piracy and counted on another war breaking out soon when they could return to legal privateering. For the professional privateer of the time, piracy was a way to make a living and stay in practice between wars. The only trick was to not get caught.

Meanwhile, Nassau had continued to grow as a privateer base and by the war's end in 1713, an estimated 200 families of settlers and 1,000 privateers called Nassau their home. Among those privateers were captains like Benjamin Hornigold, Charles Vane, and Henry Jennings, all who would become far more famous pirates over the next several years. Now with the war's end, Nassau was a pirate base. But even though Nassau was the biggest pirate base of the day, it couldn't possibly compare to the bases of Tortuga or Port Royal as they were in the 1660s. In 1713, even with a fairly large population, the port itself was comprised of only a few small buildings surrounded by a large group of run-down shacks. However, in just a few years, that would all change.

Chapter 14
The 1715 Treasure Fleet

Since the mid-1500s, all the gold and silver that was mined in Peru, Bolivia, and Ecuador was taken to Panama and all the treasure from the Spanish possessions in the Pacific was brought to Acapulco by a dozen or so Manila Galleons each year. All this treasure was then transported across the Isthmus of Panama by mule caravan to either Portobelo or Nombre de Dios. At the same time, Aztec gold along with cocoa and other valuable products were taken to Veracruz while other rich commodities from South America were taken to either Cartagena or Maracaibo. All this wealth would eventually make its way to Havana, Cuba where officials would carefully record the exact amounts and box the treasure for shipment to Spain. When all this bureaucracy was complete, the treasure would be loaded onto a single fleet of large ships known as the "treasure fleet." Once a year, this fabulously wealthy Spanish treasure fleet would sail from Cuba to Spain, timed precisely to miss the hurricane season. However, the War of the Spanish Succession changed all that. Threats from large enemy naval attacks forced Spain to stop all shipments until the war was over. But the gold, silver, and other riches continued to accumulate in Cuba on schedule.

After years of waiting, the amount of treasure in the Havana warehouses was enormous. In addition to gold and silver, there were large amounts of porcelain, ivory, and silk from the Philippines, as well as an impressive assortment of pearls, emeralds, and other jewels. By the end of the war, Spain was in desperate need of funds, so the decision to send the treasure fleet was made, regardless of the risk. The job of command went to General Don Antonio de Echevers. Normally, the fleet would sail from Havana in March, long before the hurricane season; however, this year was different. Endless delays forced the fleet to wait until the middle of the summer to sail. First, the mule trains that brought the treasure over the Isthmus of Panama were slower than usual, causing the initial sailing date to be pushed back to April 11, 1715.

The next delay was caused when a sudden storm hit Veracruz on March 28, 1715, damaging the ships that were about to sail for Havana. It took over five weeks to make repairs.

The Spanish ships assembled were the *Nuestra Señora de la Regla*, the *Santo Cristo de San Roman*, the *Nuestra Señora del Carmen*, the *Nuestra Señora de La Popa*, the *Nuestra Señora del Rosario*, the *Urca de Lima*, the *Nuestra Señora de las Nieves*, the *Maria Galante*, the *El Señora San Miguel*, the *El Cievro*, and the *Nuestra Señora de la Concepción*. An additional French ship, the *Griffon,* was added to the fleet. The *Griffon* was a 70-gun warship that had been sent to Veracruz to bring back 48,801 pieces of eight that the Spanish owed the French government for the loan of two warships in the recent war. Once all the ships were assembled at Havana, the treasure had to be recorded and boxed which pushed the departure date back even further. All this greatly annoyed General Don Antonio de Echevers but there was nothing he could do. The wealth was enormous. The lead ship alone carried 1,300 chests of coins plus gold bars, jewels, pearls, and China porcelain. The official estimation of the entire treasure carried on the fleet totaled over 15,000,000 pieces of eight. This, of course, was only the official count. In addition to the officers and crew, the treasure fleet was carrying hundreds of passengers back to Spain, many of whom were exceptionally wealthy. There is no way of knowing how much wealth was brought on board as personal property of these passengers. Also, there is no way of telling how many jewels and pearls were being smuggled aboard the treasure fleet by officials not wanting to share a percentage of the value with the crown.

1715, July

Spanish treasure fleet sinks off the Florida coast

With the King of Spain demanding his treasure, a decision was made for the fleet to sail as soon as it was ready, even though it was right in the middle of Florida's hurricane season. By the end of July 1715, the treasure fleet was finally ready to sail. On July 24, 1715, the Spanish treasure fleet left Havana and at 4 a.m. on July 31, 1715, the 11 Spanish ships were hit by a tremendous hurricane. Confined in the Florida Straits with no room to maneuver, the ships were dashed about by fierce waves and blown westward to the shallow banks. Only the French ship, *Griffon*, escaped because it was so far ahead that the hurricane completely missed the ship. The *Griffon* arrived safely in France unaware of the fate of the rest of the fleet. As for the 11 Spanish treasure ships, the hurricane drove every one of them onto the shallow Florida banks and destroyed them all.

It is believed that most of the ships went down all along the southeast Florida coast with at least one getting as far north as Amelia Island. Three of the ships sank in the shallow waters near San Sebastian Inlet and were salvageable. By the fall of 1715, the Spanish established a camp near San Sebastian Inlet and began salvage operations. Havana provided the most sophisticated

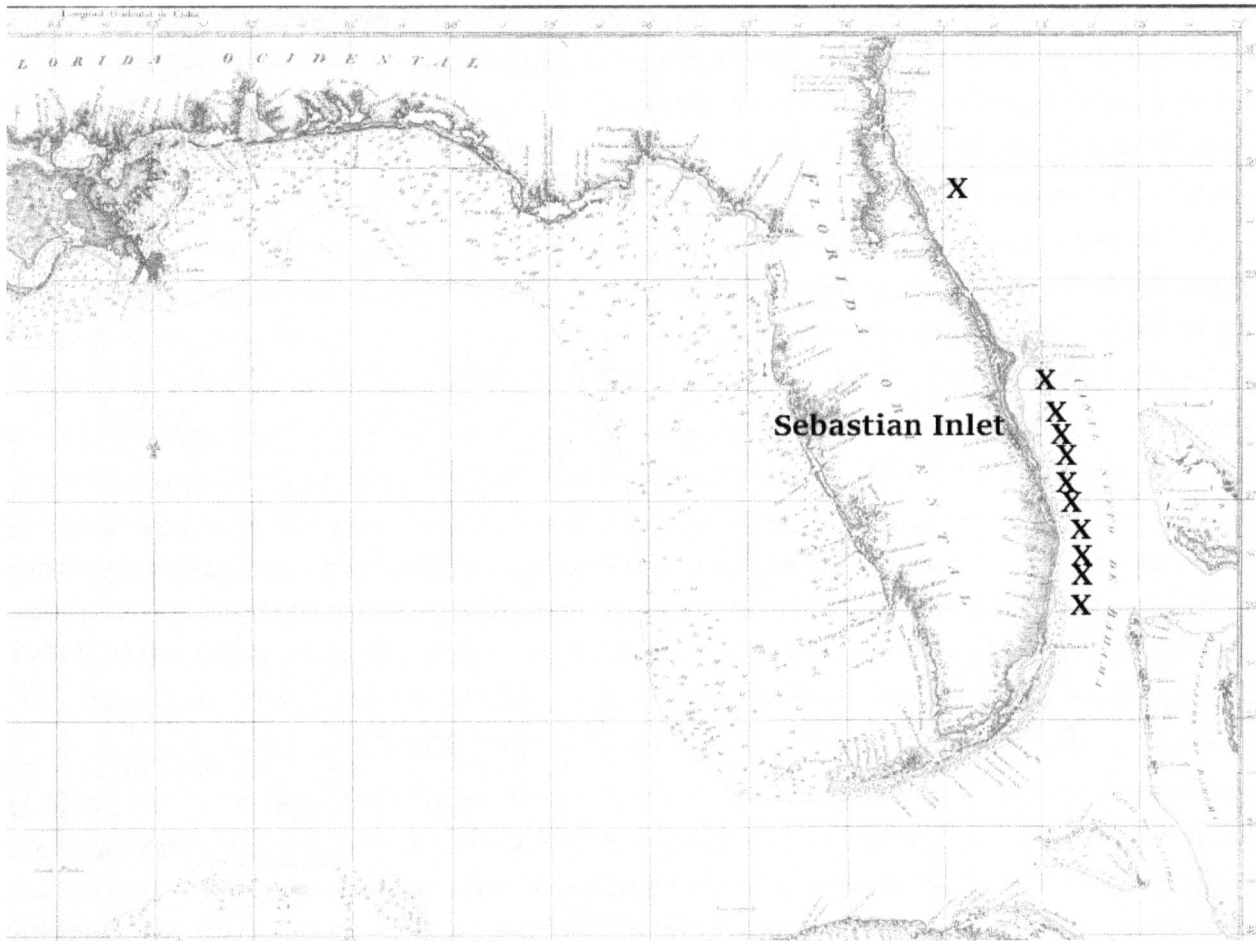

Figure 15: *Approximate Positions of the 1715 Treasure Fleet Wrecks*

salvage equipment available in the early 18th century such as diving bells. Since the three salvageable ships sank in relatively shallow water, much of the treasure from those vessels could be recovered and huge amounts of gold and silver began accumulating at the salvage camp, under the guard of only 60 Spanish soldiers.

News of the sinking of the Spanish treasure fleet spread rapidly throughout the Caribbean and the entire eastern seaboard. Hundreds of opportunists began planning ways to loot the Spanish shipwrecks. Even English government officials were supporting and encouraging the looting of the Spanish wrecks. Governor Spotswood of Virginia wrote a letter to King George I of Great Britain saying, "There is advice of considerable events in these parts that a Spanish Plate Fleet, richly laden, consisting of eleven sail . . . are cast away in the Gulf of Florida . . . I think it is my duty to inform his Majesty of this accident which may be improved to the advantage of his Majesty's subjects by encouraging them to attempt the recovery of some of the immense wealth."

The McLarty Treasure Museum, located on the site of the salvage camp at Sebastian Inlet, is an exceptionally well-done museum with many artifacts recovered from the wrecks.

Jennings attacked the salvage camps on December 27, 1715.

This sparked an international race to the Florida coast by everyone who could sail a ship or row a boat. But for most, they arrived too late. One of the former Nassau privateers by the name of Henry Jennings beat them all to it. Acting quickly, the Governor of Jamaica, Archibald Hamilton, gave Jennings secret orders on November 21, 1715, to attack the Spanish salvage camp and obtain as much treasure as he could. The reason for the secrecy was that Governor Hamilton was not planning on sending the profits he realized from the looting of the Spanish treasure fleet back to Great Britain.

Needless to say, Jennings jumped at the chance. He and his partner, John Wills, sailed their sloops to Florida. Jennings's sloop *Bersheba* had 8 guns and 80 men, and Wills's sloop *Eagle* had 12 guns and 100 men. On December 27, 1715, in the early morning light, just as the sun began to peek over the horizon, one hundred fifty well-armed men calmly walked up to the Spanish salvage camp with their weapons drawn. The Spanish must have been astonished and somewhat confused. Spain was at peace with everyone. The Commanding Officer approached Jennings and politely asked if their nations were at war again. Jennings answered no, they were there to fish the wrecks. Objections from the commander fell on deaf ears. Jennings even ignored an offer of 25,000 pieces of eight if they left immediately. The Spanish watched helplessly as Jennings's men looted the camp of 120,000 pieces of eight. On their way back to Jamaica, the pirates took a Spanish merchant ship with a large amount of silver, plus a valuable cargo of Indigo, which they sold when they returned to Jamaica. Estimations as to the value of the treasure taken vary greatly, but it is believed to be about 300,000 pieces of eight. No other raids on the salvage camp occurred. The Spanish finished their salvage operations and left Florida in April of 1716. The treasure hunters who arrived afterwards only found small amounts of loot that had either been overlooked by the Spanish or occasionally washed up on the beach.

As for the treasure fleet shipwrecks, of the 11 ships that sank, seven are believed to have been found between Fort Pierce and Sebastian Inlet and another is believed to have been found near Amelia Island, but positive identification remains elusive. There is still debate among experts as to which wrecks have been discovered and no one can say for certain where the others lie. A large amount of the treasure may still remain along the shores of southeast Florida. In 2015, a single cash of gold coins from the 1715 treasure fleet was recovered that valued $4,500,000. Modern treasure hunters are continually finding gold and silver coins just offshore or even along the beach. Additionally, fabulous pieces of jewelry occasionally turn up. Treasure hunters are especially productive after storms, where the wind and rough surf wash treasure ashore. That is why the region near Fort Pierce is called the "Treasure Coast."

Chapter 15
Nassau and Florida During the Era of The Pirates

Henry Jennings used the new-found wealth he acquired by raiding the treasure fleet salvage camp to build the town of Nassau into a new pirate base. In most 18th century source documents, this pirate base was generally referred to as New Providence. Some sources call the pirates who sailed from Nassau, "the pirates of New Providence." Both Nassau and New Providence are correct as Nassau is the name of the town located on the Island of New Providence in the Bahamas. Under Jennings's influence, Nassau quickly developed into a pirate port like Tortuga or Port Royal in the old buccaneer days. It was a port city where pirates could gather and plan ventures against all flags. Once the docks and ships' supply stores were in place, taverns and brothels were built where hundreds of pirates would spend their money. Jennings brought in prostitutes from colonies throughout the Caribbean to work in these brothels. Like most successful developers, he realized that the real profits don't come from taking prizes at sea; they came from taking the money away from the pirates who took the prizes at sea. To protect his pirate empire, Jennings rebuilt an old fort and mounted sufficient guns to thwart off any enemies. He also kept a captured Spanish ship of 32 guns in the harbor as a guard ship. Nassau had just about anything else a pirate would need. On top of that, there was no government. English authority was absolutely nonexistent. The pirates did as they pleased.

The last era of the "Golden Age of Piracy" had begun, the era of the pirates. It was by far the most significant as many of the high-profile pirates come from this era. With a new base to operate from, ex-privateers from all over the Caribbean flocked to Nassau. Pirates like Benjamin Hornigold, Sam Bellamy, Olivier Levasseur (La Buse), Charles Vane, Jack Rackham, and of course, Blackbeard. But the sinking of the 1715 treasure fleet which provided

1716

Henry Jennings establishes Nassau as a base for pirates

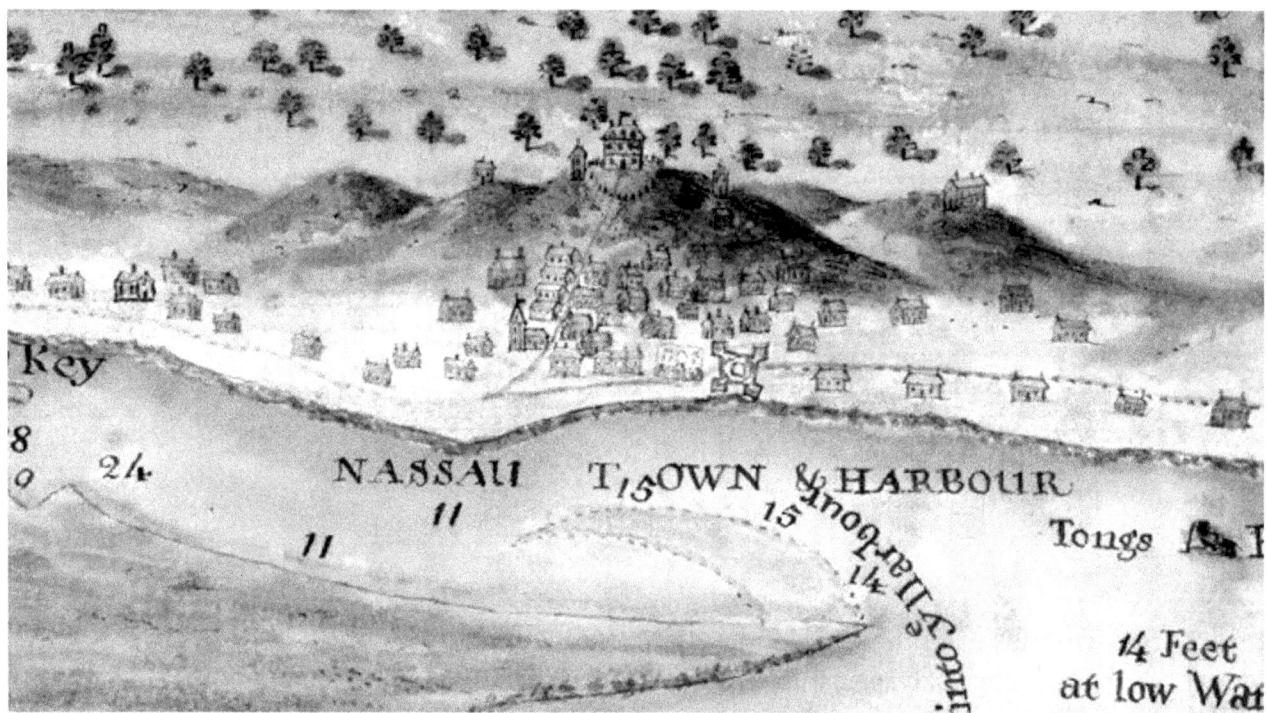

Figure 16: *Nassau*

AUGUST 1, 1714

George of Hanover became
King George I of Great Britain

the financial basis for the establishment of Nassau as a pirate base was just one of the events that sparked the era of the pirates. The other was the 1715 civil war in Great Britain over who was the rightful king, George of Hanover, or James Edward Stuart.

After Queen Anne died, the last of the Stuart rulers in England, George I became king and a civil war erupted. Many thought the rightful king was James Edward Stuart, the son of the deposed King James II and the younger brother of Queen Anne. He was living in exile in France and was denied the English crown because he was a Catholic. He was often called *The Pretender*, the *Old Pretender,* and *The Pretender Across the Sea.* Supporters of James were called Jacobites. The Jacobite revolution divided the English nobility as well as the commoners. Most of the armed conflict in this civil war took place in Scotland, but many people in England were bitterly divided by the crowning of George of Hanover as well. There is no way of telling exactly how much impact this event had on the English pirates of the time. For some individuals, it had an enormous impact, giving them a political purpose and justifying their actions.

Most historians and people in general think of all pirates in simple terms as people who simply wanted to rob others at sea for profit. But, for many of the pirates shortly after 1715, there was a clear political motivation for their piracy. Many pirates were actually Jacobite supporters. This is evident by the many names of pirate vessels that show support for James or the Stewart monarchy. Blackbeard named his ship the *Queen Anne's Revenge* in reference to Queen Anne, who was the last of the Stuart monarchs in England and the half-sister to James Edward Stuart. And Blackbeard's partner, Stede Bonnet, named his sloop the *Royal James* in direct reference to James Edward Stuart. Additionally, many pirates during this time were known to drink to the health of the *Old Pretender*, wishing to see him King of Great Britain someday. This has been reported by former captives who spent time among these pirates or by prison guards who overheard these pirates while they awaited trial.

English pirates dominated this era far more than in the past and patriotism seemed dead among most of them. English merchant vessels were being attacked at much higher rates throughout the Caribbean and along the coasts of North America and West Africa by pirates who were predominately English. Even major English seaports were being attacked and blockaded by these English pirates. Certainly, English ports and shipping had been attacked many times in the past by the Spanish, the French, and the Dutch, but they had never before been attacked by English pirates, at least not to that extent. Blackbeard attacked and blockaded ships outside the ports of Philadelphia and Charles Town. Charles Vane also attacked and blockaded shipping out of Charles Town. The Great Pirate Roberts destroyed English ports in Newfoundland and West Africa. Perhaps the struggle over the rightful King of Great Britain played a significant role. I think that it did. At the very least it may have caused some pirates to feel betrayed by their government and therefore no longer obligated to remain loyal.

Dynamic personalities among pirates also marked this era. Between 1717 and 1722 pirates seemed to be far more flamboyant and exaggerated than at any other time. Pirates used their reputation to help intimidate their prey. Each pirate captain had his own personalized pirate flag so that prize ships could identify them. The pirates also dressed to impress and to help intimidate the prize crews. Military dress with baldrics and the occasional pair of cavalry boots returned as did the eclectic mix of plundered clothing. That gave them a psychological advantage over their captives during the looting which encouraged their cooperation. The terrifying image of Blackbeard with burning matches in his thick and tangled beard is a perfect example and the Great Pirate Roberts, who dressed in the most extravagant clothing with gold chains about his neck, is another. The shocking news of female pirates Anne Bonny and Mary Read also defined this era as unique to the others. There certainly had been female pirates before and after, but these two were the first to be sensationalized in news stories and books. They captured the imagination of millions of readers throughout the American colonies and many European nations.

The media frenzy over pirates also played a gigantic role in the shaping of the "Era of the Pirates." There had been books about pirates before, but now, far more people were reading them and more books were being written every year. Newspapers were covering stories of pirate attacks for the first time ever, primarily because there were very few newspapers before the early 18th century. Songs about pirates dominated the taverns and ballrooms and the theater erupted with plays and productions based on pirate themes. It was like the gangster craze in the 1930s where gangster-themed stories dominated the media in movies, radio shows, novels, comic books, and newspapers. This media frenzy helped stylize those pirates into the characters we think of today. In many cases, the media wasn't too far off the mark.

The attention the media gave to pirates in the early 18th century also had a significant impact on public opinion. In past years, many colonists viewed pirates as bringers of exotic goods and lavish spenders of money. Pirates were good for local economies. The fact that they took these goods through piracy didn't seem to matter. They were thought of almost like we think of Robin Hood. But the stories that were written during this era turned public opinion against the pirates. They were looked upon as evil and cruel criminals. That made it easier for officials to pass and enforce anti-piracy laws which kept local merchants from doing business with pirates.

As for Florida, even though the idea of pirates operating along the Florida shores in the early to mid-18th century is intriguing, most of the pirate activity during that time was in the Caribbean or further north in the Atlantic. The prime locations for pirates were the waters of Cuba, Puerto Rico, the Lesser Antilles, and the Gulf of Honduras. That's where the wealthy plantations were located and that's where the richest prizes sailed. Spain still had no interest in constructing any new cities along Florida's east coast. They had St. Augustine, and that was enough.

Merchant ships making the long trip back to Europe and consequently traveling through the Florida Straits and along the east coast of Florida tended to travel in groups for safety, just like the treasure fleets did. It was far easier to pick off a lone vessel sailing between ports within the Caribbean than to attempt to capture convoys of merchant ships traveling through the Florida Straits. Not to say that lone ships didn't travel through the straits or that pirates didn't occasionally attack them, but it was less common.

Along Florida's west coast and throughout the Gulf of Mexico, there still weren't any settlements or ports large enough to attract pirates. Even when New Orleans was added to the list of coastal settlements in 1718, the economics in the Gulf didn't change very much. Most of Florida was in the possession of Native Americans. In the previous century, pirate raids of missions were encouraged by the governments of the English colonies of Carolina and Jamaica and therefore were somewhat profitable. But in the 18th century, that had changed. Anti-piracy laws throughout the English colonies, especially in South Carolina and Jamaica, forced pirates to seek prizes elsewhere.

Piracy slowed significantly in 1718 when Nassau's pirate operations abruptly halted when it was converted into the royal seat of the English government virtually overnight. And in the Caribbean, the ex-buccaneer port of Port Royal became a place where pirates were tried and executed. That included the famous pirates Charles Vane and Calico Jack Rackham. In 1722 Gallows Point welcomed forty-one pirates to their death in only one month. During that time, all the pirate-friendly ports along the east coast of North America literally slammed their doors on pirates.

But the final nail in the coffin for the "Golden Age of Piracy" was the deeds of the most successful pirate captain of the era. The "Great Pirate Roberts" raised the stakes. Before Roberts, pirates were merely a nuisance to the authorities. They were certainly out to stop them, but on a relatively small scale. The great powers of Europe had more important things to focus on. Plus, the pirates might be needed again as privateers if another war broke out. Roberts changed this. After 1720 his crew lost touch with reality. They tortured and killed their captives and totally destroyed many prize vessels. Roberts and his men also destroyed entire ports, one after another. This threw the entire shipping industry into a panic. Roberts moved piracy from a local problem to an international one. As a result of Roberts' attacks, almost all ports throughout the world rejected pirate trade.

Roberts was finally killed on the morning of February 10, 1722, when the *HMS Swallow* attacked his two vessels near Annobon Island on the west coast of Africa. His third vessel had been seized by the *HMS Swallow* a few days earlier. Of the 254 men who were captured and taken to Cape Corso Castle, 19 died before their trial and the 70 black pirates were all sold as slaves. Of those who were tried, one was pardoned, 20 testified against Roberts and received lighter sentences, one was found guilty but died shortly afterwards, 52 were executed, and 17 were sent back to prison in England. However, 74 were acquitted, including 16 French prisoners who were being held on Roberts' vessel. The rest consisted of all the musicians and the recent additions to the crew who convinced the court that they were forced to join. With the highly publicized capture of the Great Pirate Roberts' crew and the sensationalized death of Roberts himself, the "Golden Age of Piracy" came to a close.

Chapter 16
Did Blackbeard Visit Florida?

There is no doubt that Blackbeard is the most famous pirate of all time. He has been portrayed in more motion pictures and television programs than any other pirate and stories about his deeds are still being told and retold in books, articles, magazines, on the internet and in social media. Blackbeard's fame and notoriety in the press goes all the way back to 1718 when Blackbeard was killed. Accounts of his final battle at Ocracoke in November of that year were front-page news throughout all of North America and most of Europe. In 1719, a 12-year-old boy named Benjamin Franklin even wrote a song about the battle titled *The Downfall of Piracy*. This was the same Benjamin Franklin who signed the Declaration of Independence 57 years afterwards.

Consequently, Blackbeard's reputation has permeated local legend throughout the east coast of the United States. Dozens of communities claim that Blackbeard once operated from their shores. It's similar to the old saying "George Washington Slept Here." Just like Captain Kidd, tourists who visit Florida cities that claim to have a long pirate tradition, such as Amelia Island, will generally be told that Blackbeard operated out of that city too. Of course, they will also be told that Blackbeard must have buried some of his treasure nearby. But did this famous pirate actually stop at Amelia Island or anywhere along the Florida coast to bury treasure?

Like most of the pirates of the early 18th century, we know little about the early life of Blackbeard. Facts are hard to find and there have been so many stories written about Blackbeard that it is sometimes difficult to separate the truth from the fiction. The single source that contributed to most of the misinformation about Blackbeard is the series of editions about pirates written by Charles Johnson. The English people had become obsessed with pirates in the early years of the 18th century. Acts of piracy were front-page news. Consequently, there were lots of pirate-themed publications including

Figure 17: *Blackbeard from Johnson's 1736 Edition*

newspaper articles, fiction and nonfiction books, journals, and tabloids. There were also a great many theatrical productions that were being performed. This was propelled by the real-life deeds of many famous pirates of that time such as Blackbeard, Captain Kidd, and Calico Jack Rackham and his two female pirates, Anne Bonny and Mary Read. An unknown author named Charles Johnson wrote the most successful of all those publications, so successful that it still remains in print today.

Johnson's first edition was titled "*The General History of the Robberies and Myrders of the Most Notorious Pirates*" and was published in the spring of 1724. A second edition was published a few months later titled "*A General History of the Pirates, from Their First Rise and Settlement in the Island of Providence, to the present time. With the remarkable Actions and Adventures of the two Female Pirates, Mary Read and Anne Bonny.*" Both editions are similar regarding the content, although some of the text on Blackbeard that originally appeared in the chapter on Stede Bonnet was moved to the chapter on Blackbeard. It appears that Blackbeard became the more important of the two in the public eye. One can easily check some of the "facts" chronicled in these editions to see that they are not accurate. For example, the details of Blackbeard's final battle at Ocracoke were recorded and stored at the Admiralty Office in London as part of the official record. They were available to the public by 1721. However, in Johnson's book published three years later, his version is far different than the official report. Johnson also tells his readers about a battle between Blackbeard's ship, the *Queen Anne's Revenge* and the Royal Navy vessel, HMS *Scarborough* off Nevis Island late in 1717. A quick check of the *Scarborough's* logbook reveals that no such battle ever took place.

Johnson's first two editions, published in 1724, and all the subsequent editions published every several years until his death in 1737 and even one published in 1813 have actually served to cloud our understanding and perception of those pirates. Because the first two editions were written during the age of piracy, many historians throughout the centuries have elevated them to the status of absolute fact. They assumed that all the details described in those publications were absolutely true and accurate. Even today, dozens of modern accounts of the pirates mentioned in Johnson's books are taken almost word for word. But Charles Johnson was writing a bestseller, he wasn't writing a history book. The challenge for modern historians is that while many of the facts represented by Johnson are exaggerations, embellishments, or sensationalized fiction, many are actually correct. When researching pirates of the early 18th century Johnson's books are a good starting point, but "everything" must not be taken at face value. Many of his facts are true, but many are not.

Some modern researchers believe Johnson was actually the celebrated author, Daniel Defoe, who was fascinated with pirates and authored several fictional books on the subject, the most famous being his novel, *Robinson Crusoe*, published in 1719. But it is far more likely that Johnson was actually Nathaniel Mist, the writer and publisher of the London tabloid, *Weekly Journal*. Mist was a former sailor and was fascinated with pirates too. He was also a personal friend of Daniel Defoe. He printed over 100 stories about pirates in his tabloid between 1717 and 1724 and ran the first promotional advertisement for his book on May 9, 1724, a week before the release of the book. Additionally, the book was registered under Nathaniel Mist's name with "His Majesty's Stationers" company on June 24, 1724.

There are several different spellings of Blackbeard's actual name used in the early 18th century. "Teach" was how his name was spelled in an article that appeared in the *Boston Newsletter* in 1717. Recent research indicates that "Thache" is probably the more correct spelling based upon family records from Jamaica and England. It was probably pronounced "Tay-ch" which would account for the variations. Johnson chose the spelling of "Thatch" in his first edition and "Teach" in all subsequent editions. The 2nd edition spelling of "Teach" has influenced authors and historians ever since.

According to Johnson's 2nd edition, Edward Teach was born in Bristol and migrated to Jamaica where he became a privateer in Queen Anne's War. With the end of the war, he turned to piracy and eventually joined the pirate Benjamin Hornigold in Nassau sometime before 1716. There is some very strong evidence that Johnson may have gotten some of those facts right. Contemporary documentation from Jamaica identifies a man named Edward Thatch who was originally from Wiltshire, England and traveled to the port of Bristol with his family and then migrated to Jamaica sometime before 1698. Among his family was a son named Edward Thatch Jr. who subsequently joined the Royal Navy. Edward senior died in 1707 without leaving a will and legal documents concerning his inheritance were sent to his son who was serving aboard the HMS *Windsor*. So far, no information about Edward Thatch's whereabouts between 1707 and 1716 has been uncovered.

However, there is another very credible theory that Blackbeard actually came from Barbados and relocated to Goose Creek near Charles Town in Carolina in the late 17th century. In this version, his real name was Edward Beard, the son of James Beard, a notable ship's captain and privateer in Queen Anne's War. The family was financially successful and eventually relocated to Bath, North Carolina where James Beard purchased a substantial amount of land. He also befriended Governor Charles Eden and Tobias Knight, two figures who played a vital role in Blackbeard's later life. If this theory is correct, it would easily explain how Blackbeard was able to receive a pardon from his father's friend, Governor Eden, in 1718. According to this version, Edward Beard was a privateer in the war along with his father. I believe that the truth is a combination of both accounts. It is possible that both Edward Thatch and Edward Beard knew each other, perhaps while serving as privateers. At the war's end, Edward Thatch died at sea and when Edward Beard decided to turn to piracy, he chose to assume the identity of his dead friend as an alias to protect his sister, who remained in North Carolina.

At any rate, there is no doubt that Blackbeard was in Nassau in November of 1716 and sailing with the master privateer, Benjamin Hornigold. It is generally believed that Blackbeard was sailing with Hornigold a year or two earlier, even though there are no known surviving accounts of any pirate named Thatch or even any pirate matching his description operating anywhere in the Caribbean or Atlantic before November 1716.

Blackbeard and Hornigold operated in a counterclockwise circle around Cuba, taking 41 prizes in a single month during the spring of 1716. The first time Blackbeard entered the historical record was in December of 1716 when he and Hornigold took the brigantine *Lamb* off the western tip of Hispaniola on the 13th of that month. The ship's master, Henry Timberlake, reported a man named Thatch was the captain of a pirate crew that participated in the looting of the *Lamb* alongside the crew of Benjamin Hornigold. The following year, 1717, they swept around Cuba again, this time sailing down to the Gulf of Honduras where they took a prize near Portobelo. From there, they headed north and left the Caribbean, sailing through the Florida Straits and along the Carolina coasts, taking a Spanish ship near Charles Town. Afterwards, they headed to Accomack, Virginia, where they beached their vessels for needed repairs. It is believed that they returned to Nassau after leaving Virginia. Sometime in the late summer of 1717, Blackbeard formed a partnership with another pirate named Stede Bonnet, who had been a wealthy plantation owner on Barbados. Stede Bonnet had recently decided to become a pirate and purchased a sloop which he named *Revenge.* Afterwards, he sailed into the Caribbean. Details of how this partnership formed are highly speculative.

Traveling north in September 1717 aboard the sloop *Revenge*, Blackbeard and Bonnet took the sloop *Betty* off Cape Charles, Virginia. From there, Thatch and Bonnet sailed to Cape May, New Jersey, which is the entrance to the busy port of Philadelphia. In a bold move, they took 9 prizes in the first days of October of 1717, and Philadelphia came to a standstill. Apparently, Thatch was known to at least some of the sailors he encountered while taking their vessels as he was identified as a sailor who frequented the Philadelphia port. *The Boston News Letter* printed a story on October 24, 1717, that read, ". . . taken 12 days since off Cape by a Pirate Sloop called Revenge of 12 guns

1716 & 1717

Teach with Hornigold in the Caribbean

October 1717

Blackbeard and Bonnet take 17 prizes off Cape Charles, Cape May, and Long Island

and 150 men commanded by Teach, who formerly sailed mate out of this port. . . . On board the sloop is Major Bennet, but has no command . . . " This was the article that spelled his name "Teach."

Thatch also had acquired the nickname of "Blackbeard" sometime before his attack on Cape May. An account of the pirate's actions was recorded in a letter written on October 23, 1717, by Jonathan Dickinson, the Mayor of Philadelphia. In that letter, Dickinson names the leader of the pirates raiding vessels off Philadelphia as "One Capan Tatch All[ia]s Bla[ck]beard."

From there, he traveled to Long Island, where he took another 7 prizes. Blackbeard decided to keep several of the prize vessels he captured at Cape May. None of those vessels remained with him for more than a month. After leaving the waters of New Jersey, Blackbeard sailed east and then down to the Lesser Antilles. On November 17, 1717, while cruising 60 miles east of Martinique, Blackbeard took a slave ship named *La Concorde* and kept it, renaming it the *Queen Anne's Revenge*. This was a large three masted ship that mounted 40 guns and was built for speed. At the time, it was the most powerful pirate ship in the Caribbean.

November 1717

Blackbeard takes La Concorde and renames it the *Queen Anne's Revenge*

With the *Queen Anne's Revenge* as a flagship and the *Revenge* alongside, Blackbeard sailed north along the west side of the Lesser Antilles and then west, passing the south of Puerto Rico and Hispaniola, taking prizes as he went. That was when his alleged battle with the HMS *Scarborough* supposedly took place. After that imaginary battle, Blackbeard took the prize sloop *Margare*t and held the ship's master, Henry Bostock, for 8 hours as the pirates looted his vessel. Later, Bostock gave a detailed description of Thatch as having a large black beard.

In December of 1717, Blackbeard made repairs near Samana Cay in the Bahamas then spent the first four months of 1718 taking prizes along the west side of Yucatan to Campeche and then deep into the Gulf of Honduras. By April 1718, he had 4 vessels in his fleet, the *Queen Anne's Revenge*, the sloop *Revenge*, the sloop *Adventure*, and another small Spanish sloop taken in the Florida Straits near Cuba. In late April 1718, he returned to Nassau where he learned about the King's Proclamation offering a full pardon to all pirates who accepted it. The main requirement was that each person accepting the pardon hadn't committed any piratical acts after January 5, 1718.

Then, Blackbeard did something absolutely astonishing that has confounded historians for 300 years. He blockaded Charles Town, South Carolina for 5 days, taking 8 prizes of comparatively little monetary value. He also demanded that the governor give him a chest of medicine. While he was waiting for the medicine, many of his crew went ashore and terrorized the city. After leaving Charles Town, he sailed his fleet of 4 vessels to Topsail

May 1718

The Queen Anne's Revenge sinks at Topsail Inlet.

Inlet, North Carolina and ran the *Queen Anne's Revenge* and the *Adventure* aground, sinking them both. Stede Bonnet and his pirates were furious at the loss of their treasure and left in the *Revenge* which they renamed the *Royal James*. Meanwhile, Blackbeard and a select group of only 25 pirates sailed to Bath in the small Spanish sloop which they renamed the *Adventure*. At the time, Bath was the capital of North Carolina and the home of the Governor. Once in Bath, Blackbeard and his pirates accepted the King's Pardon from Governor Charles Eden. Afterwards, Blackbeard established a small temporary base on the island of Ocracoke.

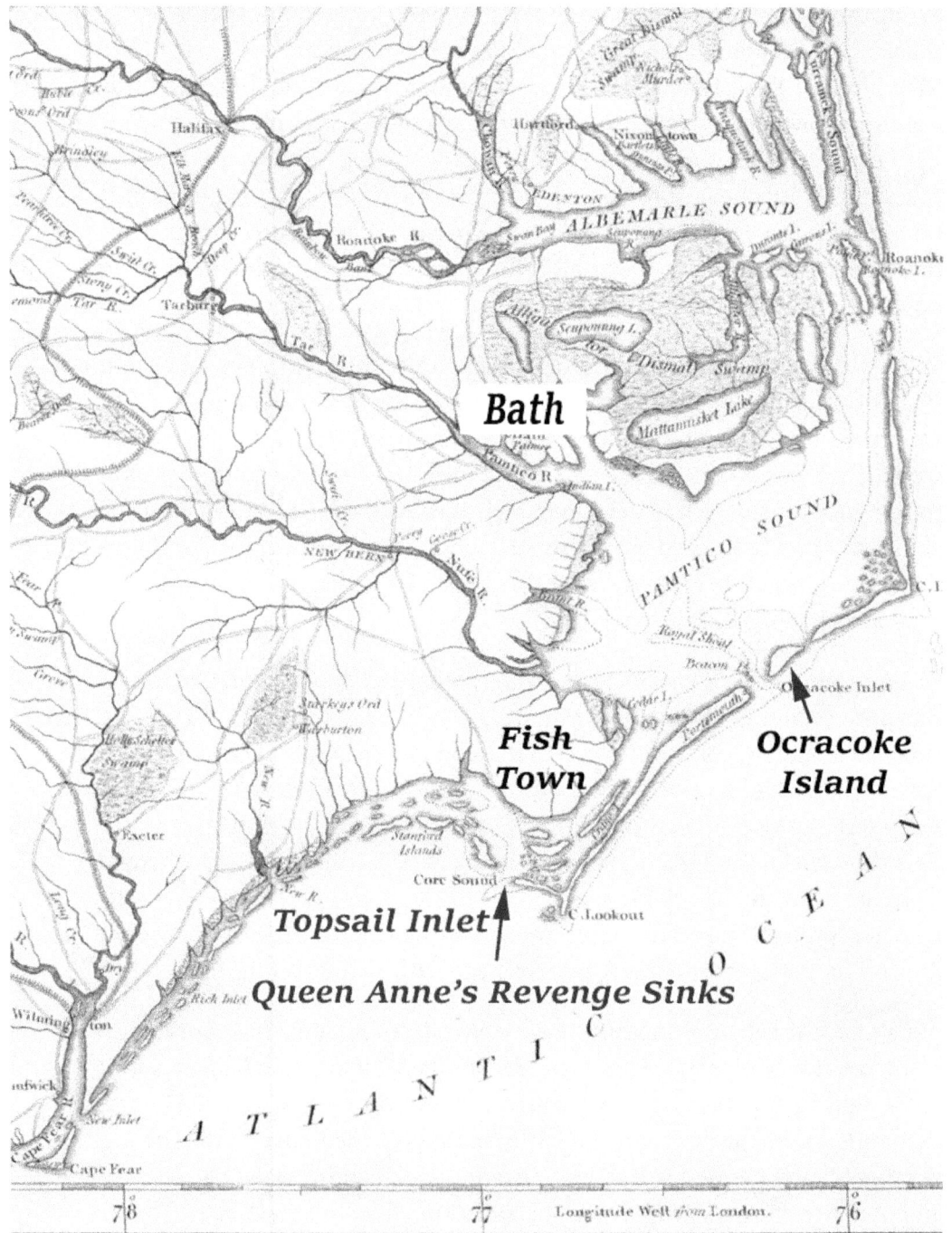

Figure 18: *Topsail Inlet and Ocracoke Island*

Stede Bonnet and his crew were captured on September 26, 1718, near Cape Fear, North Carolina and taken to Charles Town to stand trial. Bonnet and most of his men were eventually hanged. Only a few members of Bonnet's crew who testified against him were pardoned. As for Blackbeard, in August 1718, his presence in Pennsylvania near Philadelphia is documented by a warrant for his arrest issued by the Governor of the colony. It is believed that Blackbeard went to Philadelphia to visit his girlfriend Margaret, who lived in Marcus Hook just a few miles south of the city. In September, Blackbeard took two French ships near Bermuda that were carrying a large cargo of sugar. After loading all the sugar onto one of the ships, he allowed both crews to leave onboard the other ship, then took the ship with the sugar back to his base on Ocracoke.

Bath was the Capital of North Carolina and Charles Eden was the governor.

In Bath, Blackbeard met with Tobias Knight, the Customs Inspector and Chief Justice for the colony. Blackbeard told Knight that he had found the merchant ship abandoned and submitted a formal request to keep the cargo of sugar. His request was granted, but curiously, much of that cargo of sugar wound up in the possession of Governor Charles Eden and Tobias Knight.

In October of 1717, while at their Ocracoke base, Blackbeard and his crew were joined by Charles Vane and his crew. Together, the pirates enjoyed a week-long party. Blackbeard's movements during the first half of November 1718 are well documented and place him in the Pamlico River between Bath and Oriental, North Carolina. On the evening of November 21, 1718, Blackbeard and 20 others were asleep aboard the sloop *Adventure* anchored at his base on Ocracoke Island. Blackbeard's luck ran out early on the morning of November 22, 1718, when he was attacked by a British force of about 50 men aboard 2 sloops led by Lt. Robert Maynard. In the battle, Blackbeard and 11 of his men were killed. The nine others were arrested and taken to Williamsburg to stand trial.

Of the nine, six were found guilty and executed. The other three were released. Two of them apparently were not pirates, they were just visiting. The third was a black man named Black Caesar. He was released on the grounds that he was a slave and was therefore not responsible for doing what his master ordered him to do. Black Caesar was most likely the slave of one of the visitors from Bath. However, he may have been a member of the pirate crew, as five of the pirates who fought alongside of Blackbeard at Ocracoke were also black. I shall discuss Black Caesar in a chapter later on.

So, did Blackbeard operate out of Florida or even bury his treasure there? His whereabouts from late 1716 until his death in 1718 are extremely well documented. Perhaps more than any other pirate of his time. There was simply no opportunity for Blackbeard to have ever operated from a base along the Florida coast. He only passed the east coast of Florida a few times, on his way to and from Accomack in the summer of 1717, on his way to and from Cape May in late 1717, and on his way to Charles Town in May of 1718. On each of those occasions, it doesn't seem logical for him to have stopped anywhere along the Florida coast to raid shipping, especially since there are no reports of him taking any vessels anywhere along the east coast of Florida during those times. Of course, since no one knows anything about Blackbeard's actions prior to 1716, he may have used Florida as a base back then, but he wasn't known as Blackbeard yet nor was he even known to the public as a pirate. Local legends and tales about Blackbeard from before 1716 could not have been passed down through the generations because no one knew the name of Blackbeard.

But what about a short stop to bury treasure? Many locals on Amelia Island claim that Blackbeard buried some of his treasure on the island. There's a local legend that tells of a large tree with a chain hanging down from one of the branches. Supposedly, Blackbeard or some other pirates buried their treasure beneath the tree and the chain marks the exact spot. The legend continues with a highly creative ghostly tale of four men who found the tree with the chain. They began digging and eventually uncovered the top of what appeared to be a treasure chest. Three of the men went to get a bigger shovel, leaving the fourth behind to stand guard over their discovery. But curiosity got the best of the guard and he managed to dig through the dirt with his hands, clearing enough away to uncover the entire lid which he then began to pry open. As the lid of the chest slowly lifted, a black ghostly spirit suddenly arose through the opening of the chest and seized the startled guard. The mysterious spirit dragged him through the woods, eventually leaving him on the beach. When the other three returned, the tree was gone, the chain was gone, and even the hole they dug was gone. They searched for their missing friend and found him nearly dead a few hours later.

If Blackbeard wanted to bury some treasure on an island along the east coast, he could have made a short stop to bury a small chest. It would only take a few hours. But Blackbeard most likely wouldn't have stopped on Amelia Island. It was inhabited at that time, so any ship that stopped to bury a small chest would easily have been seen by the locals. Hiding a treasure chest in the midst of a local population isn't a very good idea. It would probably be gone by the time you returned for it. Additionally, there are lots of uninhabited islands along the Carolina coast that would have made a far better hiding spot. Unfortunately, I would have to say that just like Captain Kidd, Blackbeard did not visit Florida either.

Chapter 17
Calico Jack Rackham and Anne Bonny's Honeymoon in Florida

Many modern books on Florida Pirates tell a tale of Calico Jack Rackham and Anne Bonny honeymooning on a West Florida island in 1719. They certainly were real pirates. Anne Bonny and her pirate friend Mary Read rocketed to fame after their trial in 1721. Newspapers exploded with sensational and tantalizing stories of two female pirates who had been serving aboard a pirate sloop under command of Calico Jack Rackham and with a crew of 13 men. Nothing like that had ever happened before. Just think of it, two females living in sin among a crew of male pirates! Newsworthy opportunities like that just didn't get any better for an 18th century journalist. The story had everything— crime, sex, excitement, adventure, and scandal. Every newspaper throughout Western Europe and the North American colonies ran the story. Anne Bonny and Mary Read were perhaps the most famous women of their time.

It was October 1720 and Captain Calico Jack Rackham and his crew were pirating off the Jamaica coastline in their small and unimpressive sloop named the *Vanity*. Rackham already had a well-established reputation as a pirate captain and Governor Woodes Rogers had recently issued a warrant for his arrest. Pirate hunter Jonathan Barnet sighted the *Vanity* laying quietly at anchor off the north coast of Jamaica. As Barnet's men boarded the small sloop, only two members of Rackham's crew offered any resistance. Well-armed and dressed like men, Anne Bonny and Mary Read waged a fierce battle against the boarders, shooting pistols, wielding cutlasses, and spewing all sorts of profanity as they fought. Apparently, the rest of the pirates were either too drunk or too scared to offer any resistance. Bonny and Read were finally subdued and the entire pirate crew was taken back to Jamaica to stand trial for piracy.

The news spread quickly. As far as scandalous stories about sex and violence go, a story about female pirates was about as good as any newspaper could hope for. Just the idea must have shocked, thrilled, and excited readers throughout the western world. Think of what modern tabloids would do with a story like this today. It would run for weeks. Additionally, in the best traditions of tabloid journalism, the missing facts would be filled in with all sorts of sordid details that were invented by the highly creative editorial staffs. That was especially true in the early 18th century when exaggeration and embellishment by the press were the order of the day. But what is the truth? Were these women real? Where did they come from? Why did they become pirates? They certainly were real; they were captured by Jonathan Barnet in October 1720 and brought back to Port Royal for trial. But who these women were before they became pirates and how they came to be a part of Rackham's crew is far more speculative. Three hundred years later, is it possible to separate the truth from the myth?

The story of Jack Rackham and Anne Bonny's Florida adventure begins about 18 months before their capture in October 1720. In most versions that appear in books about Florida pirates that were published in the late 20th and early 21st centuries, Anne Bonny came to Nassau with her husband James in 1719. Shortly afterwards, she met the charming Jack Rackham, who had recently accepted the king's pardon and was living in Nassau. Jack began to court her and offered her some jewelry. James wasn't much of a man and Anne began a relationship with Jack. Her outraged husband dragged Anne to the Governor Woodes Rogers, and demanded that she be flogged for her outrageous behavior. A court order was issued forbidding the two lovers from seeing each other. But Jack and Anne were in love, so Jack decided to return to piracy and leave Nassau with Anne as his new partner. In the summer of 1719, Jack assembled a small crew which consisted of some of his former pirate crewmen and a few new ones, which included Mary Read, who was a woman disguised as a man.

Needing a pirate vessel, Jack and his pirates set their sights on a fast merchant sloop named the *Curlew*, which was anchored in Nassau's harbor. In the darkness of night, as Rackham and his pirates approached the two men on anchor watch, Anne Bonny raised her blunderbuss and threatened to blow out their brains if they resisted. This establishes the narrative of Anne being highly aggressive and forceful, which is actually supported by historical accounts given by their victims. Jack and the rest of the crew quickly boarded their new pirate sloop and set sail for Cuban waters to hunt for prize ships. Over the next few weeks, they were very successful, taking small prizes such as fishing vessels and small trading sloops. This success changed while sailing off of Key West when they attempted to take a large Spanish sloop. Instead of surrendering, the Spanish sloop opened fire. A well-placed shot toppled the main mast of the pirate sloop *Curlew* and Rackham had to break off the attack.

Captain Jack Rackham sailed his crippled sloop northward in search of a tree tall enough to replace his mast. Unfortunately, they ran into a tropical storm that blew the *Curlew* further north into the Gulf of Mexico. The crippled vessel limped into safe anchorage off the southern end of Estero Island near present-day Fort Myers, Florida. While the pirates affected repairs, Jack and Anne spent the time honeymooning on a nearby island named "Lover's Key." Some modern works suggest that the island got its romantic name from the local legend of Jack and Anne's honeymoon. There is only one original source of this story which says, ". . . while the ship's carpenter and crew did labour mightily to refit the vessel, Calico Jack and the woman, Anne Bonny, did make merry for many days on the island where they had a dwelling made of sticks and palms."

The earliest source I could find on this version of the saga of Jack Rackham and Anne Bonney in Florida is the book, *Pirates and Buried Treasure On Florida Islands* by Jack Beater, published in 1959 by St. Petersburg: Great Outdoors. Subsequent publications had repeated this story, citing Beater's book as their primary source. Beater cites his source of Jack and Anne's adventure in Florida as Charles Johnson's *History of Highwaymen and Pyrates*, published in 1818 by DuPlau of London. If Charles Johnson's name sounds familiar, it is. This is the same Charles Johnson who wrote *A General History of the Robberies and Myrders of the Most Notorious Pirates* in 1724. As mentioned in the previous chapter, Charles Johnson was most likely Nathaniel Mist, the writer and publisher of the London tabloid, *Weekly Journal*, that specialized in stories about pirates. But how can this be? Charles Johnson, or Nathaniel Mist, died in 1737, many years before 1818. If true, this edition was simply rewritten by the publisher who owned the printing rights. Johnson's original book and all subsequent editions were filled with unsubstantiated stories and facts written to sensationalize the book and to increase sales. You may recall that Johnson, or Mist, was not writing a history book, he was writing a bestseller.

Initially, everything we know about the lives of Anne Bonny and Mary Read before 1720 comes from Charles Johnson's early editions published in 1724. Both tell the same story; however, the 2nd edition highlights their names in the title, *"A General History of the Pirates, from Their First Rise and Settlement in the Island of Providence, to the present time. With the remarkable Actions and Adventures of the two Female Pirates, Mary Read and Anne Bonny."* Apparently, the 1818 edition adds more tales and even changes some of the events from the earlier editions. If we assume that the facts about Anne Bonny, Mary Read, and Jack Rackham are as embellished, sensationalized, and even fictionalized as we have seen with Blackbeard and even other pirates in that book, then we must realize that we really know nothing about them for certain. So far, researchers have not been able to find a single document, record, or even contemporary account that mentions anything about the lives of Anne Bonny or Mary Read before September 1720. Over the centuries, the life stories of these two very real women have been told and retold hundreds of times by authors who have contributed to the invention of their mythical pasts. Their true stories remain cloaked in mystery. So far, Charles Johnson's books are the only sources of their earlier lives. His version might be completely true, or it might be partially true, or it might be a complete fabrication. However, there is no doubt that Jack Rackham played a vital role in the lives of Anne Bonny and Mary Read.

Unlike Anne Bonny and Mary Read, the information we have on Jack Rackham before 1720 is well documented and verified by many sources. Calico Jack Rackham got his colorful name because he supposedly preferred to wear cotton clothes. Don't we all when traveling in the Caribbean? In 1720 the word "calico" didn't refer to the same brightly colored cloth it does today. In the early 18th century, most of the cotton found in the colonies was imported directly from Calicut, India by the famous English East India Company or indirectly through French merchants. Because all cotton came from Calicut, even plain white cotton, it was commonly known as calico.

Calicut was a major trading port in India beginning in the 13th century. Today, that city is named Kozhikode, India.

Rackham was the quartermaster for the famous pirate Charles Vane when he shot his way out of Nassau in the summer of 1718 and when he took between eight and twelve vessels at the entrance of the Charles Town Harbor. He was still Vane's quartermaster when Vane and his crew spent a week partying with Blackbeard's crew on Ocracoke Island in October 1718. Operating in the Windward Passage between Cuba and Florida on November 24, 1718, Jack Rackham led the mutiny that removed Vane from the position of Captain. Vane and a few of his loyal crewmembers were permitted to leave and Calico Jack Rackham was elected the new captain.

Rackham enjoyed moderate success taking small prizes close to the Jamaican shoreline. In December 1718 he took his first rich prize, the merchant ship *Kingston*. The problem was that he captured the prize within sight of the harbor of Port Royal, Jamaica. This infuriated the local merchants who hired several pirate hunters to track down Rackham. In February 1719 the pirate hunters finally found Rackham's brigantine *Ranger* lying at anchor along the recently captured prize ship *Kingston* off Isla de Pinos, just south of Cuba. While Rackham and his entire crew were asleep on the beach in tents that were made from old sailcloth, the pirate hunters approached the vessels. Rackham and his crew became alerted and scrambled into the jungle and eventually escaped on foot. Afterwards, he obtained a small boat somewhere on the island and sailed away with six of his crewmen.

For the next three months, Rackham stayed at sea, traveling around Cuba, and eventually making his way to Nassau. Calico Jack Rackham must have been an exceptionally persuasive talker because he managed to convince Governor Woodes Rogers that he had been forced into piracy by Charles Vane. Eventually, Rogers granted Rackham and his crew the King's Pardon.

The Myth of Anne Bonny

According to the account written by Charles Johnson, Anne Bonny was born in Kinsale, Ireland in the county of Cork just before 1700. She was the illegitimate daughter of a prominent lawyer named William Cormac and his housemaid. Due to the scandal of the relationship, the father, the maid, and Anne left Ireland and settled in Charles Town, South Carolina. William Cormac eventually became a wealthy plantation owner, but his family life soon turned to tragedy. At age 13, his daughter, Anne, stabbed a servant girl with a table knife. Then, a few years later, she married an apparently unimpressive sailor named James Bonny.

The two newlyweds wound up in Nassau just as Governor Woodes Rogers was issuing pardons and stamping out piracy. James Bonny assisted Rogers by informing him about the activities of those who remained pirates while Anne's love interests wandered. In the late spring of 1719, the very handsome and dashing Calico Jack Rackham arrived in port soon after he had escaped capture off the coast of Cuba. The two of them fell in love. When the husband protested the relationship, Rackham offered to pay James Bonny what was called a divorce for purchase, but the stubborn and money-conscious Anne wouldn't allow it. The two of them stole a sloop and headed out to sea. For Rackham, it was his chance to return to piracy and for Anne, it was the beginning of an adventure. Their sloop was renamed the *Vanity* and Captain Rackham began recruiting a crew directly from other pirate vessels at sea. At first, Anne attempted to disguise herself as a man, but soon gave up the pretense. Eventually, Anne became pregnant with Jack's child and had to give up pirating just long enough to give birth somewhere in Cuba. She left her child with her Cuban friends and returned to sea as a pirate.

The Myth of Mary Read

According to the account written by Charles Johnson, Mary was from Portsmouth, England and was born around 1690. Her father was a sailor who was lost at sea. With the family funds running out, Mary's mother went to her mother-in-law for financial assistance. Mary's mother figured there was a better chance of getting support if she had a son, so she dressed Mary as a boy, and they maintained the deception even after the mother-in-law's death. As a teenager, Mary sought adventure. Keeping the male disguise, she joined the Royal Navy and served aboard a man o' war. Fearing her true identity would be discovered, she changed jobs often.

Mary left the navy and enlisted first in an infantry regiment and then in a cavalry regiment in Flanders. Mary eventually fell in love with a soldier who knew her true gender and they were married. Mary began dressing in woman's clothing for the first time in many years, but when her husband died, she returned to her old lifestyle. Mary signed on as a sailor onboard a Dutch merchant ship which brought her to the Caribbean. The ship was attacked by English pirates and Mary joined their crew, keeping her true gender a secret. After Governor Rogers offered the King's Pardon, Mary and the pirate crew that she had joined decided to accept the pardon and become privateers. It was 1719 and there was a war on with Spain. Mary was sailing in this capacity when the vessel she was on was overtaken by Captain Jack Rackham, looking to recruit pirates for his crew. Mary figured that the opportunity for wealth was greater as a pirate, so she joined Rackham's crew, still maintaining her disguise.

In the beginning, Anne Bonny was the only one onboard who realized that Mary was indeed a woman. They became very close friends. Most versions go a lot further than that and suggest that they were actually lovers. This is where the story really gets exciting. In many versions, Rackham entered the ship's cabin and found Anne and Mary in the midst of making love. Still thinking Mary to be a man, Rackham became enraged and commenced to attack Mary. Standing between the two, Anne quickly revealed the true nature of Mary's sex, which rapidly changed the dynamics of the entire situation. Afterwards, Anne and Jack maintained their love relationship, Anne and Mary maintained their love relationship, and it seems that Mary began a love relationship with one of the male crewmembers, to whom she revealed her true gender. One can easily understand how a tale like this generated a lot of interest when it was released in print. This was the sex story of the century.

Back to the Facts

Exactly how and when Anne Bonny and Mary Read came to be onboard Jack Rackham's sloop is unknown and will probably always remain a mystery. We know that Rackham was in Nassau in the spring of 1719 after he and six of his crew escaped the Isla de Pinos. We also know that neither Anne Bonny nor Mary Read were with Rackham before this. Perhaps Anne Bonny did meet Rackham in Nassau exactly as told. It seems likely. In any event, on August 22, 1720, Jack Rackham returned to piracy and stole a sloop named the *William*

AUGUST 1720

Calico Jack Rackham cruises off Cuba

which he renamed the *Vanity*. In early September 1720 near Harbor Island, Bahamas, he took several small prizes, mostly fishing vessels, and increased the size of his crew by recruiting directly from other pirate vessels he encountered. With a full crew, he returned to his old stomping grounds, *or should I say stomping waters,* near the shores of Cuba and Jamaica. News of Rackham's piracy reached Nassau later that month. Governor Woodes Rogers nullified the King's Pardon he had recently given to Rackham and issued a warrant for his arrest. Over in Jamaica, Governor Nicholas Lawes sent pirate hunter Jonathan Barnet in a well-armed sloop to seek out and capture Rackham and his crew.

Meanwhile, Rackham was cruising off Hispaniola on October 1, 1720, when his crew fired warning shots at two passing merchantmen. The ships surrendered and the pirates boarded. Anne Bonny and Mary Read were mentioned for the first time in the accounts later given to the authorities by the victims. The women were described as screaming like banshees and frightening all the members of the prize crew. A short time later, while cruising the waters off the north coastline of Jamaica, Rackham's sloop stopped a canoe near the shore. The canoe had only one occupant, a woman named Dorothy Thomas. In her testimony at the trial of Rackham's crew, she stated that "the Two Women were on the Board the said Sloop, and wore Men's Jackets, and long Trousers, and Handkerchiefs tied about their Heads; and that each of them had a Machet and a Pistol in their Hands and cursed and swore at the men to murther (me)."

Then on October 19, 1720, Rackham and his crew took a schooner captained by Thomas Spenlow near Port Maria Bay on the northern coast of Jamaica. According to Spenlow's testimony during Bonny and Read's trial, the women were the first to board and continued to intimidate and terrorize the members of the prize crew as they went about taking clothing and anything else they desired.

Sailing west along the north coast of Jamaica, Rackham reached Dry Harbor Bay on October 20, 1720, where he hailed the merchant sloop *Mary*, commanded by Thomas Dillon. Upon seeing Rackham's sloop *Vanity*, most of the crewmen jumped overboard and swam ashore. Several warning shots were fired, and Captain Dillon surrendered. The *Mary* was robbed of about £300 in cargo. At the trial of Anne Bonny and Mary Read, part of Dillon's testimony reads "Anne Bonny had a Gun in her Hand, they were both very profligate, cursing and swearing much, and very ready and willing to do most anything on Board."

Early on the evening of October 22, 1720, the *Vanity* lay at anchor in Dry Harbor Bay. As their sloop gently tugged at the anchor cable, the pirates

OCTOBER 22, 1720

Calico Jack Rackham, Mary Read, and Anne Bonny are captured.

were below deck, playing cards and drinking rum. At about 10 o'clock that night, Anne and Mary were on deck when they sighted another sloop headed directly towards them. The two women shouted out a warning and Rackham and a few others scrambled up the ladder to the main deck while the rest of the pirate crewmen remained passed out below. It was the well-armed sloop commanded by pirate hunter Jonathan Barnet. As Barnet approached, he gave them a hail, asking their identity. Rackham replied, "John Rackham from Cuba." Barnet ordered them to surrender peacefully, and Rackham answered by firing a swivel gun. Barnet's sloop returned fire and the male members of Rackham's crew, who were on the main deck, all dove for cover.

Realizing that he had no chance, Rackham asked for quarter and Barnet granted it. As Barnet's sloop came alongside and his men boarded the *Vanity*, the two women pirates suddenly sprang into action. According to Barnet's report, ". . . the women screamed, fighting like hellcats as they shot their pistols and swung their cutlasses, refusing to give up peacefully." On the other hand, Captain Rackham and his male crewmembers gave up without resistance. Eventually, the two women were subdued, and Rackham and his pirate crew were taken to St. Jago de la Vega, (old Spanish Town) Jamaica where they were imprisoned.

Governor Nicholas Lawes personally presided over the trial that was held on the 16th and 17th of November 1720. Two of the pirates, John Besneck and Peter Cornelian, claimed they had been forced into piracy and testified against Rackham and his crew. When it came to Anne Bonny and Mary Read, Besneck and Cornelian both stated that the women were always the first to board a prize and that they wore men's clothing while fighting but women's clothing at other times. Captain Jack Rackham, quartermaster Richard Corner, and sailing master George Fetherston were hanged at Gallows Point on November 18, 1720. James Dobbin, John Davies, John Howell, Patrick Carty, Thomas Earl, Noah Harwood and two others were hanged over the next two days. Rackham's body was placed on display as a warning, hung from a gibbet on a sandbar at the entrance of Port Royal Harbor. Today, this tiny strip of sand is known as Rackham's Cay.

SIR NICHOLAS LAWES

Governor of Jamaica from 1718 to 1722

In a surprising finish, both Anne Bonny and Mary Read revealed that they were pregnant and "plead their bellies" at their trial ten days after Rackham's execution. They were found guilty of piracy but couldn't be executed until after the birth of their children since executing a pregnant woman would be executing the child too. Charles Johnson wrote that Mary Read died in prison. According to the Parish records for St. Catherine, Jamaica, Mary Read died on April 28, 1721, and was buried in the church graveyard. This would have been near the time she was due to go into labor. Pirates weren't normally buried on sacred ground and the fact that she was buried by the church causes one to speculate that she may have died with her unborn child still inside of her. The church would have wanted to bury the child on sacred ground and if the child was still inside of Mary, they would have to be buried together. As for Anne Bonny, she too may have been buried in the same graveyard. Parish records also show a woman named Ann Bonny as having died and buried on December 29, 1733. If this is indeed the same woman, it indicates that she was released from prison and lived a normal life in Jamaica for 12 years.

As for the sexually charged relationships that transpired between Anne Bonny, Mary Read, and Jack Rackham aboard the sloop *Vanity*, we may never know the truth. It is a fact that Anne Bonny and Mary Read were both several months pregnant at the time of their trial. There are no surviving eyewitness accounts of how and when Anne Bonny and Mary Read came to be on Jack Rackham's sloop, but it is an absolute fact that Anne Bonny and Mary Read were on board the *Vanity*, dressed as men and fiercely fighting to prevent capture when they were arrested by Jonathan Barnet in October 1720.

Back to Florida

The big question about Jack Rackham and Anne Bonny's adventure in Florida is how did this story surface, almost 100 years after it supposedly occurred? The Florida honeymoon between Jack and Anne did not appear in any of the earlier editions that were published just a few years after this event allegedly took place. In the early 18th century, Johnson's books were a product of the public's fascination with piracy. However, this fascination began to wane by the mid-18th century and public interests turned to other topics. This changed in the early 19th century when piracy in the Caribbean literally exploded as revolutionary governments throughout South America and Mexico chose to use pirates to help wage war and raise funds. The dramatic increase in piracy triggered a renewed interest in stories about pirates in the public mind.

Charles Johnson's editions had been out of print for decades and the few surviving copies were selling in London for large sums of money. The time seemed right for a new edition and on January 1st, 1813 *The History of the Lives and Actions of the Most Famous Highwaymen, Street-Robbers, to which is added a genuine account of The Voyages and Plunders of the most noted Pirates,* was published by John

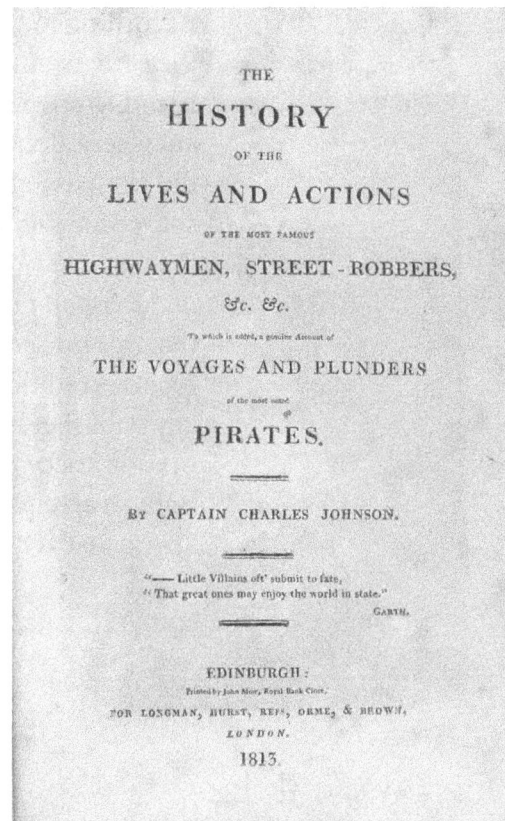

Figure 19: *The History of the Lives and Actions of the Most Famous Highwaymen, Street Robbers, etc. by Captain Charles Johnson, 1813*

73

Moir, Royal Bank Close, for Longman, Hurst, Rees, Orme, & Brown, London. However, there is absolutely no mention of the Florida honeymoon in this edition. The 1813 edition is very similar to the original 1724. It modernizes the grammar and spelling a little, but doesn't add any additional content.

According to Jack Beater's *Pirates and Buried Treasure On Florida Islands*, there is an 1818 edition that contains the story about their honeymoon in Florida. When compared to earlier editions, the name of Rackham's vessel changes from *Vanity* to *Curlew*. Additionally, the parting of Anne from her husband James is dramatized. Instead of a civil parting with Jack offering to pay for a divorce, James wants Anne flogged for her outrageous behavior and the governor issues a court order to keep Anne and Jack apart. The 1818 version also abandoned the earlier idea that Anne Bonny's true identity as a woman was at first only known to Jack Rackham. The new version has them honeymooning in a hut on an island while the crew repairs their sloop. Obviously, the crew knew her true sex. However, this story completely overlooks the fact that credible documentation places Jack Rackham in the Bahamas taking prizes when his Florida honeymoon supposedly took place.

What about the local legend that the name "Lover's Key" came from Jack and Anne's honeymoon? The island was only accessible by boat until 1965. According to the Lover's Key State Park website, the island got its name in the early twentieth century, as many young couples would row over to the island to spend some time in seclusion. And in 1719, the only inhabitants anywhere along the lower west coast of Florida were Native Americans who did not pass along any local legends of pirates. As mentioned earlier, the source for this story came from Jack Beater's 1959 book, *Pirates and Buried Treasure On Florida Islands*, cited as Charles Johnson's *History of Highwaymen and Pyrates*, published in 1818 by DuPlau of London. What is surprising is that this 1818 edition doesn't seem to exist. After an exhaustive search, I have found no edition of any 1818 publication of Johnson's book, nor could I find any publisher named DuPlau of London. It appears that Jack Beater either incorrectly quoted some other unidentified source that had invented this honeymoon story, or Beater himself invented the entire story to add to his book.

Chapter 18
Florida's Black Caesar

One of the more intriguing tales of Florida pirates is the story of Black Caesar. What makes it so intriguing is that Black Caesar seems to be either two people with the same name and operating base or one person who lived to be over 130 years old. As you may recall, Black Caesar was the name of one of the men serving aboard Blackbeard's sloop when it was captured at Ocracoke. He was arrested along with Blackbeard's other pirates who survived the battle and later released on the grounds that he was a slave and not responsible for his actions.

In one version of the legend of Florida's Black Caesar, he was a tribal chief who had been enslaved and was being transported to the colonies sometime in the early 18th century. His slave ship sank in a hurricane and Caesar and one other person managed to swim to the shore of Elliott's Key, located along the Florida Straits. Caesar and his partner signaled a passing vessel for help, but when that vessel made anchor to rescue them, they captured it, taking everything of value. They continued taking vessels using that method until they eventually recruited a pirate crew, built a small pirate base, and began using small vessels to take prizes offshore. At some point, Caesar decided to join Blackbeard's crew and left his tiny pirate haven. From there, some versions have him being hung in Williamsburg after he was caught in 1718 and other versions have him returning to Elliott's Key and continuing his piratical ways living to the 1820s.

1801–1804

Haitian Slave Revolt

The other version places him about 100 years later. In this version, he was Henri Caesar and was a slave on the French colony of San Domingue. In 1804, there was a slave revolt throughout the colony and the government was overthrown. A new nation emerged, run by former slaves. The new nation was named Haiti. As part of the revolution, Henri managed to capture a large vessel and decided to turn to piracy. By 1805, he settled at Elliott's Key which he used as his base.

Of the two basic versions mentioned, the origin of the version where Henri Caesar arrived at Elliott's Key around 1805 seems to predate the origin of the version where he arrived at Elliott's Key around 1700. It's easy to see at a casual glance how pirate enthusiasts heard the 1805 version and muddled the name of Black Caesar with the man captured at Ocracoke in 1718 as part of Blackbeard's crew. The version where he arrived at Elliott's Key in the early 1700s seems to have been recently invented in an attempt to merge both individuals into one person and force the story to fit. Since I don't believe that Black Caesar could have begun his career around 1700 and still have been sailing as a pirate 120 years later, I will close the book on that version and continue with the rest of his story.

Florida's Black Caesar established a small pirate base at Elliott's Key, on the southeast tip of Florida south of Miami. Today, it's part of Biscayne National Park. It was strategically located along the Florida Straits, the main trade route for just about every ship leaving the Caribbean. Henri chose the colorful name of Black Caesar and became one of the most successful pirates of his day. It is rumored that he accumulated treasure between $2,000,000 and $6,000,000. All this treasure was buried on several islands throughout South Florida to include Pine Island, White Horse Key, Marco Island, and Sanibel Island. Of course, to date, none of his treasure has ever been found.

But no pirate legend would be complete without tales of cruelty to captives and sexually charged tales of how the female prisoners were dealt with. According to the legend, Black Caesar routinely tortured his male prisoners, killing many of them. As for the females, he kept them all captive and forced them to serve in his pirate brothel. And the legend goes on, telling us that

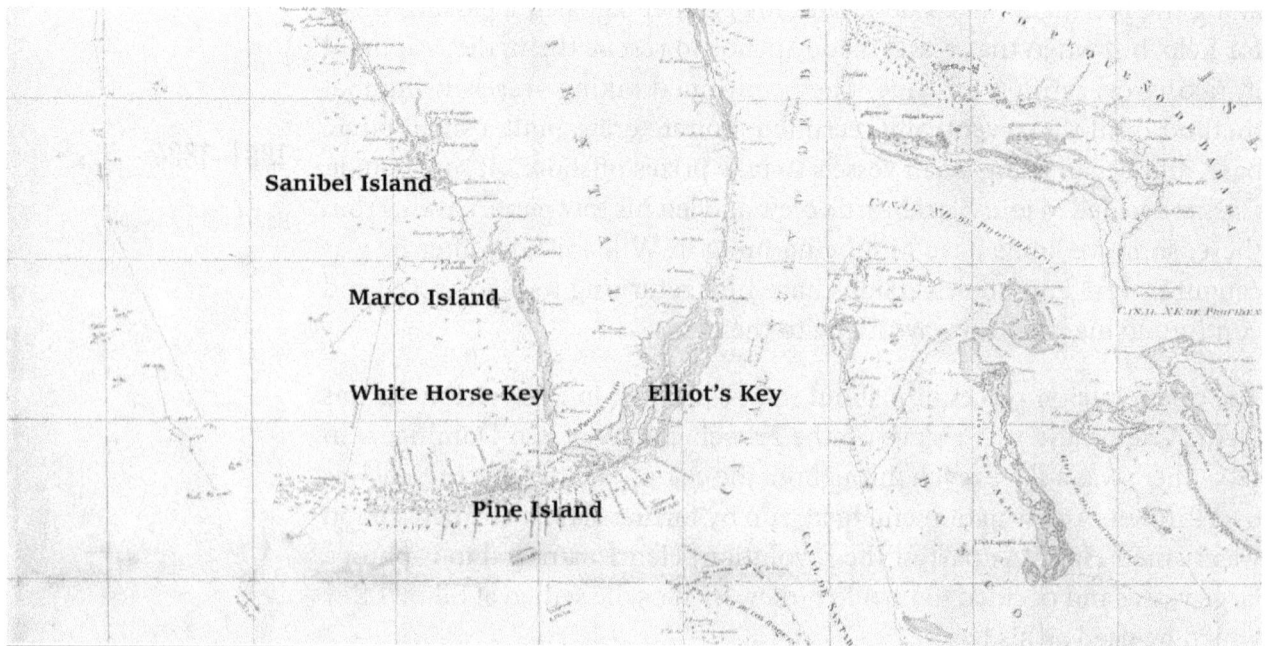

Figure 20: *Black Caesar's South Florida Treasure Spots*

Black Caesar eventually formed a partnership with the legendary Florida pirate, José Gaspar. Sometime around 1818, the two of them joined forces. Black Caesar left his base on Elliott's Key and relocated to the west coast of Florida so he could sail alongside of his new partner.

While sailing with Gaspar, Black Caesar established a residence on Sanibel Island. Gaspar kept all his loot buried on Boca Grande while Black Caesar tended to bury his loot on several of the nearby islands and throughout the keys. Black Caesar was with Gaspar when his pirate kingdom was growing. Not long after Black Caesar joined Gaspar, others like Brewster Baker and "Old King John" joined their group. But in 1819 when the treaty to make Florida part of the United States was signed, many of Gaspar's pirates began to question whether they should remain on the Florida west coast. Realizing that the United States would not tolerate a large pirate base operating in Florida, Gaspar's entire organization called for a meeting which was held on Sanibel Island. After a long discussion, most of Gaspar's pirates voted to break up the group. That included Henri "Black" Caesar. After parting company with José Gaspar, Black Caesar returned to Elliott's Key and continued his piracy just as before. It is uncertain how or even when he died.

In truth, all versions about the life of Henri "Black" Caesar are fiction. There isn't a single piece of historical documentation that proves the Florida pirate Black Caesar ever existed. The only factual documentation of any pirate named Black Caesar was the one associated with Blackbeard. The earliest reference to any pirate named Black Caesar operating in Florida came from Albert Payson Terhune, an early 20th century novelist and the author of the book, *"Black Caesar's Clan,"* published in 1921. His mystery novel takes place in a fictional tract of marsh and mangroves near Miami and features a savage pirate family. Apparently, locals began inventing their own stories about Black Caesar after the book's publication. These stories circulated around the Miami area and eventually they were told to tourists. Tourism erupted in Miami in the early 20th century, mostly spurred on by sport fishing. The guides who told the best stories attracted the majority of the clients. Locals in the shops, restaurants, and hotels also profited by telling exaggerated tales of pirates to tourists, just as today. Thus, the legend of Henri "Black" Caesar was born. Unfortunately for pirate enthusiasts, he never actually existed.

Chapter 19
Where Have All the Pirates Gone?

There must have been small-time pirates operating along the coasts of Florida in the mid-18th century. After all, pirates had been raiding Florida for decades. The Spanish government was expanding coastal settlements in Florida to counter expansion from the French in Louisiana and the English in Carolina. New Spanish settlements were springing up all along the Florida coasts, especially along the panhandle. Shipping was on the rise too, as those new settlements needed goods and supplies. But large-scale pirate activity around the waters of Florida seems to have been almost nonexistent.

In the 1720s, the British government felt that it was time to stamp out piracy in the Caribbean. They used many former pirates like Benjamin Hornigold, to serve as pirate hunters in order to help the navy drive the pirates out. The enormous amount of pirate hunters sailing from Nassau, Charles Town, and Port Royal forced most pirates in the Caribbean and the Gulf of Mexico to look for prizes elsewhere. Some pirates chose to take vessels off the waters of New England while others chose the West Coast of Africa or the Indian Ocean. From the late 1720s until the end of the 18th century, there are no famous pirates mentioned by contemporary authors or even by authors and historians in later decades. Small-time pirates must have still been operating, but no pirates with the fame and reputation of men like Blackbeard or Charles Vane.

The real reason why piracy along the Florida coasts became so limited was the expansion of the British and French colonies. That expansion brought military troops to protect the new territory and a strong military presence tends to discourage piracy. From the 1720s to the 1750s, Florida was the center of most military operations in North America. With naval forces sailing the waters and armed troops stationed at all the settlements, large-scale piracy just wasn't feasible.

The British had attempted to invade Florida several times in the early 1700s. Now with the well-established colony of Carolina, gradual expansion southward to Florida was the next step. Meanwhile, the French were pushing westward from their settlement at Fort Conde, located in present-day Mobile, Alabama. They had attacked Pensacola in 1718 during the previous war and remained a threat. But it was the British who took overt action to expand towards Florida. During the 1720s, the modern state of Georgia was a no man's land, claimed by Spain but not occupied by anyone except Native American tribes. In 1721, the British built Fort King George near the coast in the middle of that territory, about 60 miles north of the Florida border. Then, in 1733, the British officially claimed Georgia as a new colony and built the port city of Savannah. Expanding southward, in 1736 the British imported dozens of Scottish families and established several settlements in Georgia closer to Florida. The Scots were known to be both good fighters and good farmers. They would serve as a strong military unit to attack the Spanish in Florida while developing and settling the land between Florida and Carolina.

A series of formal and informal wars erupted between Great Britain and Spain during the mid-18th century. In North America until 1755, those wars were mostly fought in Georgia and northeast Florida. The final war, which began in 1756, raised the stakes considerably. It was The Seven Years' War, which started as a small conflict between Britain and France over control of the Ohio River in Pennsylvania, then quickly escalated to a major conflict that involved most of the European nations. Spain managed to stay out of the war until the last year, when they joined France against their old enemy, the British. However, France finally lost and in 1763 many French and Spanish territories changed hands in accordance with the new treaty. All of France's territory in the north, known as New France, was ceded to Britain and renamed Canada. Britain also finally achieved another objective and took control of Florida. But in a separate treaty, France turned control of Louisiana over to Spain to help pay their large war debt. At last, Britain finally controlled all of Florida and all the Spanish residents were relocated to Cuba or other Spanish colonies.

However, England's possession of Florida was short-lived. The American Revolution (1775–1783) changed everything. During the war, the British colony of Florida was the only colony south of Canada to remain loyal to Great Britain. Continental revolutionary forces invaded Florida and took control of Amelia Island for a short time. At the war's end, control of Florida went back to Spain as a reward for the Spanish support of the United States.

1718–1720

War of the Quadruple Alliance

————————

1727–1729

Anglo Spanish War

————————

1739–1748

War of Jenkins' Ear

————————

1756–1763

Seven Years' War

————————

The Spanish residents returned and took control of their major cities, St. Augustine, Pensacola, Amelia Island, and San Marcos de Apalachee. For 75 years, Florida's shores were relatively safe from pirates. The constant military operations along Florida made large-scale piracy impractical. But that was about to change.

Chapter 20
The Pirates of the Muskogee Nation

The most unexpected pirate of the Florida coasts was actually a British soldier, a businessman, a Creek warrior, an adventurer, a British Indian Agent, and the political leader of a new nation comprised of several Native American tribes. His name was William Augustus Bowles and he was perhaps the most interesting person of his day. What makes him so intriguing was the incredibly complex nature of his politics, goals, and positions. At one time or another, Bowles represented the British government, the Creek Nation, a fur trading conglomerate, and the Nation of Muskogee, which he created. But he also led a fairly large group of pirates that virtually controlled the Gulf of Mexico for almost three years. His life is as complex as his politics. If Hollywood made a totally accurate movie about his life and adventures, the critics would call it highly improbable fantasy. Yet it would all be true.

Born in 1761 in Frederick, Maryland, his father had fought in the French and Indian War. In 1775 when the Revolutionary War was ramping up, the Bowles family remained loyal to King George. When the fighting started, the 11-year-old William traveled to Philadelphia and joined a Maryland Tory Regiment. He was promoted to ensign, which is the most junior officer in the army. During the first years of the war, he served at the battle of Monmouth and was eventually stationed in New York with the occupation army.

Life quickly changed for William in December of 1778 when his regiment was sent to Pensacola, Florida. It was part of the British plan to drive into Georgia from the south. Shortly after his arrival, William got into a dispute with his superior officer and left the regiment. Under the rules of provincial regiments, officers were permitted to resign, so there were no charges or

APRIL 19, 1775

The first battle of the American Revolution at Lexington and Concord

Tories were American colonists who remained loyal to the British Crown during the American Revolution. Many of them formed Tory Regiments and fought for the British.

81

disciplinary action taken. But now, William found himself far away from home with no job or money. Eventually, William joined the Lower Creeks, who lived in a village near present-day Tallahassee. Over the next two years, he married the daughter of the chief of the Creeks and also a Cherokee princess. He produced several children from both marriages. He also established himself as a Creek Warrior, learning several native languages and becoming totally immersed in their culture.

Spain had managed to stay out of the conflict between the Continental Army and the British government until 1780. Realizing a chance to get Florida back, Spain finally joined forces with the revolutionary army. Striking from Spanish Louisiana, they launched an invasion into west Florida and attacked Pensacola. The commander of the small British force remaining in the city knew he would need help from the Creeks if he was going to defend the city. The British contacted William Bowles and asked him to organize a defensive force of Creeks. Bowles accepted and rejoined the British army. During the defense of the city, William and his detachment were stationed at Fort George, a small gun emplacement overlooking the city. Unfortunately for the British, they couldn't hold, and the Spanish forces retook the city. The British were allowed to leave as long as they promised not to fight in Florida again. When the British evacuated, William went with them, leaving his new Native American family behind.

The British troops returned to New York and remained there until the end of the war in 1783. In accordance with the treaty, all the British troops had to leave the brand-new nation called the United States of America. That included all the Tory regiments. But the Tories were all born in America and weren't really welcome in England. Each one was given a choice as to where they would now live. They could either go to Nova Scotia in Canada or they could go to the Bahamas. William chose the warmer climate, as most of the other Tories did. As an officer, William was given half pay and 500 acres on Eleuthera Island.

Lord John Murray, 4th Earl of Dunmore, was the governor of the Bahamas from 1787 to 1796.

A few sources claim that William was an actor or even a stand-up comic on the stage when he traveled to the Bahamas, but this isn't precise. While he was stationed in New York, he joined a group of other British officers who would put on stage plays from time to time. That wasn't unusual for British officers of the day, especially when they were trying to pass the time while occupying a city with no combat engagements. Other sources state that William became an artist specializing in portraits while he was in the Bahamas. That is true, but it was more of a hobby, not his only way of making a living.

At some point, William met the Governor of the Bahamas, Lord Dunmore. They not only became close friends; they became business partners. Dunmore was a very important person within British politics. He had been the Royal Governor of New York and the Royal Governor of Virginia before he was forced to leave at the start of the American Revolution. Plus, he still had some serious connections back in London. Because of his importance and his political connections, Lord Dunmore was the ideal business partner.

Meanwhile, after 20 years, Spain had retaken possession of Florida and was busy reestablishing their cities and building their economy. Spain's main concern now was expansion from the newly established government of the United States. It seems as if the settlers in Georgia, the Carolinas, and Virginia were continually moving west into the territory that Spain claimed, such as the territory that eventually became Alabama, Mississippi, and Arkansas. Some of that territory belonged to the Creeks and Cherokee, but some of it belonged to Spain. After a century of costly wars, the Spanish didn't have the money or resources to prevent that expansion. They decided to solve this problem by using the Native American tribes to keep the American expansion under control. In order to accomplish this, the Spanish had to give them enough guns and ammunition to fight the Americans; but again, they didn't have the resources. However, in an ironic turn of events, the British did. The Spanish sought help from the British; but it wasn't the British government, it was a select organization of British merchants.

The fur trade was big business in Europe. Fur was not only used to make coats; it was turned into felt which was used to make hats. The best source for fur was the Native American tribes. Spain didn't want the British merchants to become too powerful or for their trading practices to get out of control, so they decided to license only one British company to trade in Florida. That company was the Panton, Leslie & Company, which established trading posts near San Marcos de Apalachee in the panhandle and on the St. Johns River on the east coast of Florida. The company gave the Spanish a percentage of the profits and began trading all sorts of goods, including guns and gunpowder, to the various tribes located throughout Florida. But their main customers were the Creeks in northwestern Florida and Alabama. This trade immediately began to turn a huge profit.

Back on Nassau, Lord Dunmore wanted to get in on the action. He devised a plan to take the trade away from the Panton, Leslie & Company and start his own company. In order to ensure success, he needed a partner who knew the Creeks and had the skills to pull off a takeover. William Bowles, a Creek Warrior and son-in-law to two powerful chiefs, was the man. Dunmore wasn't planning on taking the business away from his competitors through shrewd deals or negotiations, he planned on taking over their operations by military force. William was promoted to the rank of Colonel and given some unofficial troops with orders to take possession of one of the trading companies. In 1788, William's small force of troops landed on the east coast of Florida and attacked the trading post on the St. Johns River. However, the attack failed and William returned to Nassau in defeat.

The following year, 1789, William and Dunmore traveled to London accompanied by a delegation of Creek and Cherokee chiefs, who were William's friends. They were all treated exceptionally well, wined and dined by various members of Parliament, and housed in luxurious accommodations. During these meetings, William and several members of Parliament devised a very ambitious plan. William would go back and lead a Creek uprising and seize Florida completely. Meanwhile, British Indian Agents in Canada would enable many tribes in the western territory, today the states of Indiana, Michigan, Ohio, and Illinois, to resist all American expansion. Once they were in control and organized, they would link up with William's Creek Nation in Florida and take control of all the land between the United States and the Mississippi River. The next step was to officially make that territory a British Protectorate. This would restrict the United States to the states along the east coast. From there, the British Protectorate could expand westward, perhaps even to the Pacific. Well, I did say that it was an ambitious plan.

In 1791, William traveled to Canada to finalize plans with General John Graves Simcoe, Lt. Governor of Canada. From there, William returned to Florida and organized a force of about 300 Creeks and attacked the Panton, Leslie & Company warehouse near San Marcos and captured it. Then he went on to capture the Spanish fort of San Marcos itself. After his success, the Spanish government invited him to travel to New Orleans, the Spanish capital of Louisiana and western Florida, in order to negotiate a treaty. From there, he was sent to Havana, Cuba to speak with the overall governor of all Spanish territory. William was exceptionally well-treated and was convinced that the negotiations were going well. He was told that the leaders back in Spain were the only ones with the authority to really make deals, so William boarded a Spanish ship in 1792 and sailed to Madrid.

However, when William arrived in Spain, everything changed. He was now a prisoner. He was still fairly well-treated but remained under house arrest. The Spanish didn't know what to do with him. They couldn't let him go back and start a revolution, but they also couldn't execute him, as they were trying to get on good terms with the British. He remained in Spain for two and one-half years, in the same state of limbo. Out of sight, out of mind, occurred to the Spanish, so in 1795 William was sent to Manilla in the Spanish colony of the Philippines. Williams' luck took a turn for the worse in 1796, when Anglo-Spanish negotiations broke down and the two countries

Figure 21: *Portrait of William Augustus Bowles painted in London in 1797*

1796–1802

Anglo-Spanish War between Britain and Spain

Mauritius Island, near Madagascar, remained neutral and serviced both British and Spanish ships during the war.

Muskogee is the actual name of the Creeks. In 1690 an English trading post was established along a small creek in the center of their territory, located in the modern city of Macon, Georgia. Merchants along the coast who shipped supplies to that post began to say that they were sending goods to the "creek." Eventually, the Muskogee tribe began to be referred to as the "Creeks" by the English.

went to war. The Spanish government sent for William, perhaps to execute him, or perhaps just to use him politically in some undetermined way to gain advantage over the British during their peace negotiations. But William never reached Spain. While his ship docked at Mauritius Island in the Indian Ocean to take on supplies, William climbed through a porthole and swam to a nearby British ship.

William returned to England where he had his portrait painted. In meeting again with members of Parliament, William was delighted to learn that the plan was still on. Only this time, since Britain and Spain were at war, support would be greatly increased. This time, William would have financial backing. This time, William would have weapons and ammunition. This time, William would have pirates.

William returned to Florida in 1799 and began organizing his revolution to take Florida away from Spain. He formed a new nation, named the Nation of Muskogee and was elected the Director General by the Creeks. He often went by his Creek name, Eastajoca. The new nation had its own flag. As soon as his government was in place, William began to form his army which was comprised of Creeks, Seminoles, Miccosukee, and escaped slaves. In May 1799, William and his new army laid siege to San Marcos and captured a supply ship, the *Sheerwater*, sent by Panton himself. But eventually, the Spanish counterattack forced William and his small army to retreat. It was a good start, but this new army desperately needed weapons, supplies, and money. That's where the pirates came in.

The British government officially recognized the Nation of Muskogee as legitimate with the ability to issue letters of marque. Privateers who sailed under those letters of marque could take Spanish, French, and even United States ships and legally sell their goods in British ports such as Nassau. Under the terms of these letters of marque, the captains kept two-thirds of the profits, which was far better than the standard rate for privateers of the time. In the late 18th century, privateers usually only received one half of the profits. British privateers from Nassau flocked to Florida to join Bowles' navy.

It is a misconception that Bowles' pirates were a rag-tag group of Native Americans and amateurs. These pirates were the cream of the crop. Highly trained professional sailors and privateers mostly from Nassau, their vessels were usually smaller sloops and schooners, not large warships. Their strength fluctuated from month to month, with as few as 3 vessels to as many as 300 vessels at any one time. Records are exceptionally scarce, so no one really knows for certain. Most of the captains were English, but there were probably a few captains from other nations as well. The crews were a mix of British, Creeks, Seminoles, Miccosukee, escaped slaves, British loyalists from the United States, and even a few Spanish.

The Admiral in charge of all the privateers was Captain Richard Powers from Nassau and his vessel was named the *Muskogee Micco*. Among other senior captains were Captain James Ward and Captain Gibson. The name of another privateer vessel mentioned by Bowles himself was the *Tostonoke*. Sailors in the Muskogee Navy operated under strict regulations, very similar to the British Royal Navy. Under these rules, to speak against the nation or to show any disrespect to the flag was a flogging offense.

So, were they pirates or privateers? One might think that they were privateers since they had letters of marque. That's what the British called them. But technically, since the Nation of Muskogee wasn't recognized by any other nation as a legitimate state, their letters of marque were also considered illegitimate. Technically, they were pirates.

1799–1802

Pirates of William Bowles control the Gulf of Mexico

From late in 1799 to March of 1802, the pirates of William Bowles controlled the entire Gulf of Mexico. Hundreds of ships were captured. During that time, Spanish supply ships were constantly sailing to and from Pensacola and San Marcos (modern-day St. Marks). Additionally, rich merchant vessels belonging to the Panton, Leslie & Company were constantly sailing to their factory near San Marcos. Bowles' pirates would strike from behind the hundreds of tiny islands located along the coast. After the prize ships were looted, the pirates would sail to Apalachicola Bay where they would meet with members of Bowles' staff. There, they would exchange any needed supplies they took, such as weapons and food, for gold or silver coins. From there,

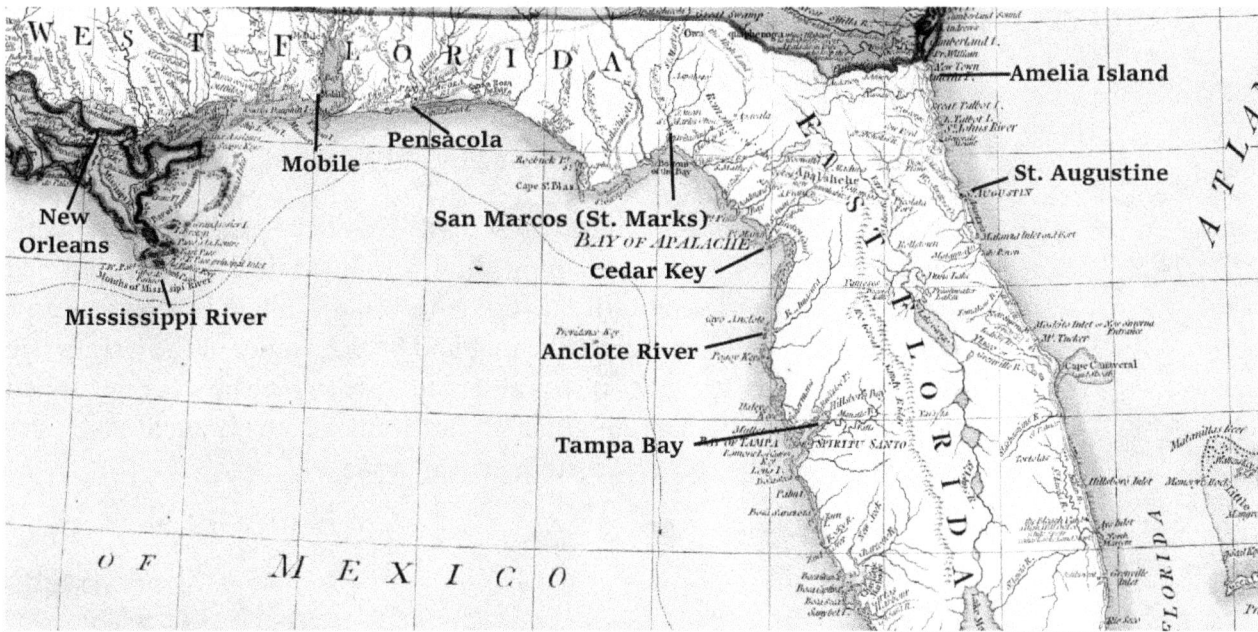

Figure 22: *Florida, 1794 Map*

the privateers would return to Nassau to spend their money. Occasionally, the British would hire these pirates to bring supplies such as weapons and uniforms directly to Bowles. In addition to the Apalachicola Bay, there were two other locations where Bowles' forces continually received supplies. One was in Tampa Bay, near the mouth of the Hillsborough River. It was the site of an old Cuban fishing village and was located near present-day Davis Island. The other spot was near Cedar Key, where Bowles' forces constructed a tall wooden lookout tower. The location of that tower was where the current lighthouse now stands.

Cedar Key seems to be the spot that Bowles' pirates used the most. The islands near Cedar Key offered great concealment which made them the ideal location to wait for passing vessels traveling along the coast and then to suddenly sail out and seize them. But one might ask, why would vessels be traveling along the coast in the first place? The answer is fresh water. There was a freshwater spring at the mouth of the Anclote River that vessels had been using to refresh their water supply for decades. The Anclote River was located about halfway between the Keys and the ports of the north coast of the Gulf. It was the perfect spot to make a short stop to take on fresh water. The location of the spring is on the site of the present-day Anclote River Park, even though there is no remaining trace of it. Any vessel leaving theAnclote River and traveling to San Marcos would have to sail right past Cedar Key. So perfect was Cedar Key as a pirate ambush spot, it would be used for that purpose a few years later by the pirates of Jean Laffite.

By January of 1802, William was ready to try to recapture San Marcos. This time, William requested artillery which the British supplied through the Muskogee base at Tampa. Just as things were looking good for William, on March 26, 1802, Spain and Britain signed a peace treaty. The war was off and Britain pulled all their support for the Nation of Muskogee. Their letters of marque were no longer recognized by Britain. Not wanting to be tried as pirates, all of Bowles' pirate crews quickly abandoned him.

With his navy gone and all British support halted, it wasn't long before his army fell apart. Bowles' remaining allies continued operating in a limited capacity for over a year until May 24, 1793, when William Bowles was captured by a Native American tribe that had been enemies of the Creeks. He was handed over to the Spanish and sent to Cuba, where he was imprisoned in Morro Castle. William Augustus Bowles, leader of the Nation of Muskogee and the Muskogee pirates, died in prison in 1805.

Chapter 21
Jean Laffite, Master Pirate, Master Spy

Without question, Jean Laffite is the most famous pirate in the history of the United States. This charismatic and bold pirate was the most successful pirate anywhere in North America during the second decade of the 19th century. But he was also an American patriot. Laffite was partly responsible for Andrew Jackson's success against the British invasion force of New Orleans in January 1815. Historians know a great deal about Laffite's actions from 1812 to 1819, but like so many other pirates, his early life is clouded in mystery and remains the subject of debate among modern historians and pirate enthusiasts. There are two main reasons for much of this confusion.

The first reason for the confusion is a book, *The Journal of Jean Laffite*, supposedly written by Jean Laffite himself and first published in 1948. This journal was introduced into the historical record by John Andrechyne Laflin who claimed to have been Jean Laffite's great-grandson. According to Laflin, he found the journal among his family papers, translated it from French to English, and made it available to the world. This single source is the major contributor to the confusion over Jean Laffite's origins.

In this journal, Jean's parents were Sephardic Jews, originally from Spain, they moved to France and then on to Port-au-Prince, San Domingue just before Jean was born. Jean grew up in a large family, but especially spent time with his older brothers, Alexander and Pierre. Alexander, who was much older, had been in the navy and was an expert naval gunner and navigator. The three of them traveled to France in 1802 and eventually wound up in New Orleans by 1805.

Jean and Alexander established a privateer operation at Barataria while their brother Pierre opened a blacksmith shop in New Orleans to be used as a front

The Journal of Jean Laffite was published in 1948

1799–1802

Cedar Key used as a pirate ambush spot by the pirates of William Bowles for their pirate and smuggling operations. Around 1812, as their operations and fleet grew, Alexander changed his name to Dominique You to avoid identification by the authorities. Jean was an American patriot who hated the British. As the War of 1812 commenced, Jean's privateers began engaging British war ships. He sided with Jackson at the Battle of New Orleans in 1815 and later relocated to Galveston where he established a new pirate base. The journal also mentions José Gasparilla, stating that he was ". . . killed by his own men on November 15, 1820." After leaving Galveston in 1820, Jean Laffite lived in many cities throughout the United States, raised a large family, and lived until 1850.

There are many problems with this work. The paper of the original journal seems to be from the mid-19th century, but the handwriting appears to match the handwriting of the journal's discoverer, John Andrechyne Laflin. Even though there are many facts and names in the journal that match the historical record, there are several names and accounts that do not. In the journal, Jean fought a naval engagement with a British warship off the south Louisiana coast in 1813. The British warship was defeated and withdrew with heavy losses. However, there is no mention of such an engagement in British maritime records. Another issue with the journal is its account of José Gasparilla, who many historians believed to be a real person in 1948 but is now generally classified as a fictional character. The final nail in the coffin for this journal comes from the fact that John Andrechyne Laflin had previously been accused of forging documents from Abraham Lincoln, Andrew Jackson, and David Crockett. I do not give any credence to this journal.

The second reason for the confusion of Laffite's early history and why it is so difficult to correctly identify the real Jean Laffite is that there appears to have been several different privateers operating in the Caribbean between 1804 and 1810 with the same name. Apparently, the name of Jean Laffite was a common name among the French just as John Smith is common among the English. Those names weren't always spelled the same way, but variations in the spelling of a name was common in that time period. When reading an early 19th century account of a privateer captain named Jean Laffite, one can't be certain if it refers to the same Jean Laffite who established a pirate base in Barataria Bay years later. Although detailed information on the early life of Jean Laffite remains elusive, far more precise information on the early life of his brother Pierre is well known. By tracing the life of Pierre Laffite, we can narrow down the truth about the early life of Jean.

Some books and references use the spelling of "Lafitte," with one f and two t's while other references use the spelling of "Laffite," with two f's and one t. In his own signature, Jean used the second spelling "Jean Laffite" and most modern historians agree that this is the correct spelling of his name.

1789–1796

The French Revolution

Early Years of the Laffites

The Laffite family lived in Pauillac, in the Bordeaux region of France. That's where Pierre and Jean were born. Most likely, their father was a high-class smuggler, bringing contraband goods over the Pyrenees from Spain. By the late 19th century, the French Revolution had destroyed the economy of France, and Pierre and Jean decided to relocate to the French Caribbean colony of San Domingue. Pierre set up business as a merchant, importing and exporting goods to and from colonies throughout the Caribbean. What Jean did is unknown, but most likely he became a privateer, eventually rising to the rank of captain.

1796–1815

Napoleonic Wars

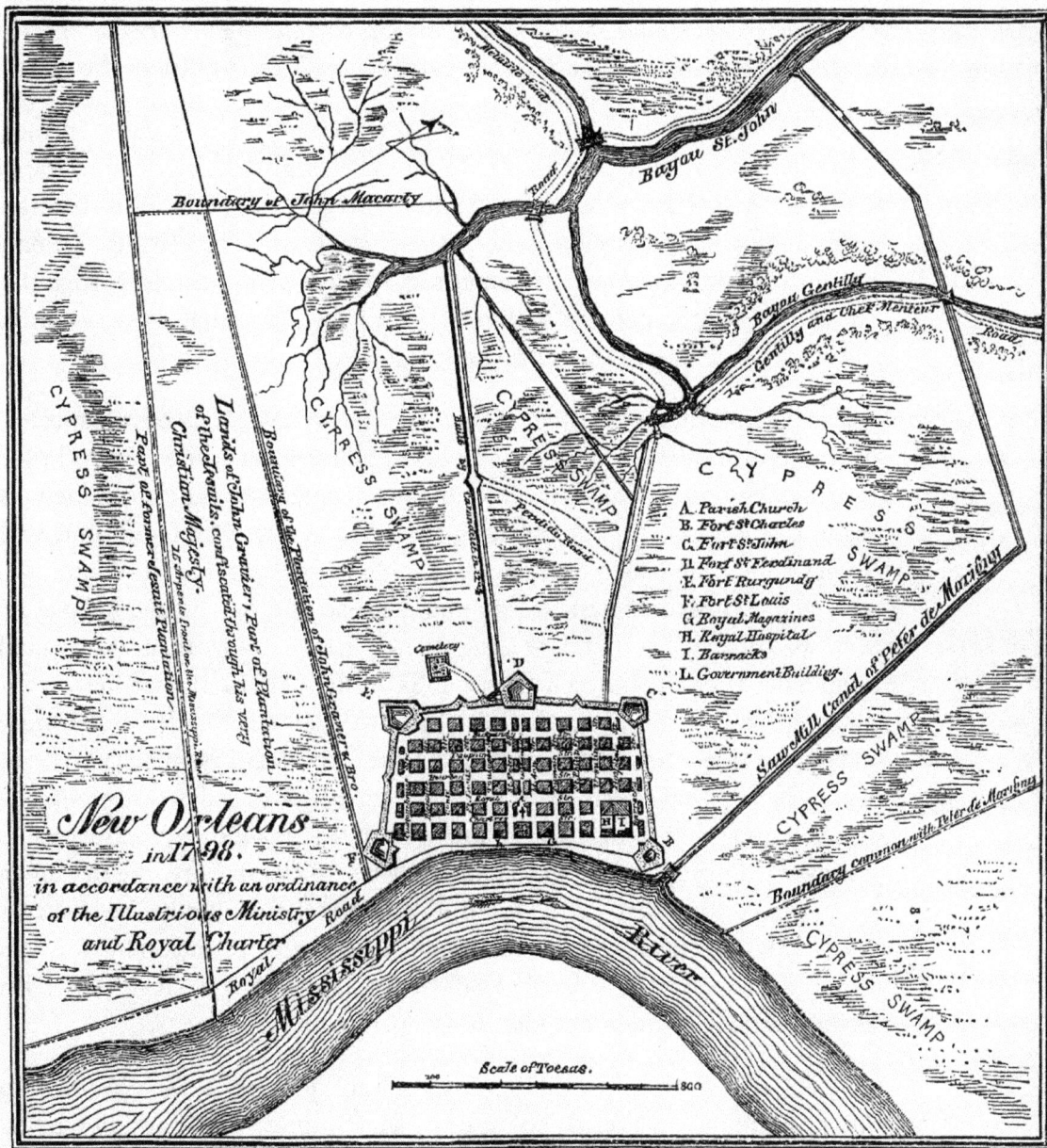

Figure 23: *New Orleans 1798 Map*

On December 20, 1803, the United States officially took control of Louisiana and William C. Claiborne assumed the position of Governor.

Beginning in 1801, the slaves on San Domingue began a series of revolts. Eventually, they took over the government and declared it to be the new nation of Haiti. During those revolts, many of the former slave owners were executed, so thousands of whites and free blacks fled the island. Most went to nearby Cuba or New Orleans. Pierre was among those who fled to New Orleans, arriving in March of 1803. After a series of failed attempts at establishing an import business, he relocated to Baton Rouge in 1804 and then to Pensacola in 1805.

New Orleans was the place to be in 1804. The United States had just acquired Louisiana from France as part of the Louisiana Purchase. Under the previous French administration, foreign trade had been severely restricted. But under the new American administration, foreign trade was encouraged and the city of New Orleans quickly became the fastest growing city in the world.

The first mention of Jean in association with New Orleans came in 1804. The details were recorded in a letter from Governor Claiborne to Pierre Lausat, dated April 1804. In that letter, the governor tells of a "Captain Lafette" and his privateer vessel *La Soeur Cherie*, who entered New Orleans accompanied by two recently captured Spanish prize vessels. Notice that the spelling of the name is with one "f" and two "t's." That was a common alternative spelling of the name Laffite.

Apparently, Captain Lafette had managed to evade the customs inspectors at the mouth of the Mississippi by taking backwater routes around their station. When he arrived in New Orleans, Captain "Lafette" told the governor that he was a privateer sailing out of Aux Cayes, San Domingue with letters of marque issued by General Jean Baptiste Brunet of San Domingue to attack British and Spanish shipping. He also said that he had been at sea since October 1803. That date is significant, because a week after he sailed, France and Spain signed a treaty, making his letters of marque invalid against Spanish shipping. Captain "Lafette" claimed that he didn't know about the treaty and thought that his capture of the Spanish prizes was legal. He requested permission from the governor to sell his cargo in New Orleans and to make needed repairs to his vessel. Claiborne was suspicious but granted him permission to land anyway. Once on shore, the privateers quickly sold their cargo and repaired their vessel. Eventually, the governor found out that "Lafette" had evaded the customs inspectors on his way into port and that one of the prize vessels was actually an American vessel. But before Claiborne could act, the elusive privateer was gone.

In 1804, after France lost their colony of San Domingue, the governors of the remaining two French colonies of Martinique and Guadalupe began issuing letters of marque to everyone who asked. It is believed that Jean

Laffite was among those privateers operating from Martinique. During that time, reports of privateer action sprang up throughout the West Indies. In those reports, there were numerous mentions of a privateer named Dominique You and his vessel *La Superbe*. There is also a report of a Captain Lafitte and his vessel *Le Regulateur* who took an American vessel named the *Maria Mischief* in 1805. That Lafitte may or may not be the same Jean Lafitte who settled in New Orleans in 1809.

The status quo rapidly changed throughout the entire Caribbean in 1808. Back in Europe, Napoleon placed his brother Joseph on the throne of Spain and the legitimate Spanish government went into exile and joined forces with the British government. Amidst the turmoil created by the change within the Spanish government, many of the Spanish colonies in Mexico and Central and South America saw the opportunity to revolt against Spanish rule and form their own governments. The most successful of those revolts was the one led by Simón Bolívar in Colombia and Venezuela. Bolívar established Cartagena as his capital and began issuing letters of marque against Spanish shipping to literally everyone who asked. Consequently, piracy dramatically increased everywhere in the Caribbean and in the Gulf of Mexico. Additionally, since France and Spain were now at war, all the French refugees from San Domingue who had fled to Cuba were expelled. They fled to the nearest French speaking city, New Orleans.

In 1809, the time seemed right to the Laffites to build a pirate base near New Orleans. First of all, letters of marque from Cartagena were easily obtainable which gave their operations a sense of legitimacy. Secondly, there was a large amount of available manpower in New Orleans. The French refugees who recently arrived from Cuba were out of work and desperate to make a living doing anything. But there was a third reason. The merchants in New Orleans were in great need of goods to sell. The United States had placed an embargo on foreign trade back in 1806, and even though the restrictions were loosened, imported luxury goods were still severely limited. To make things worse, profits from the new sugar, cotton, and rice plantations were skyrocketing and the rich plantation owners had lots of cash to spend. The merchants of New Orleans just didn't have the goods they needed. But privateers did.

That embargo was issued back on November 21, 1806, after the French changed their rules of engagement for privateers. France was fighting a bitter war with Britain as part of the Napoleonic wars. The British had won a few naval engagements and the French were determined to stop them. In 1806, the French issued new rules that allowed all French privateers to attack British shipping in ports of neutral nations such as the United States. Those rules also allowed French privateers to attack all ships from neutral nations

1808

Napoleon places his brother Joseph on the throne of Spain and the Spanish government in exile changed alliances from France to Britain.

1808–1821

Simón Bolívar led a revolution against Spanish rule that involved the current nations of Venezuela, Bolivia, Colombia, Ecuador, Peru, and Panama which were originally united as the nation of Gran Colombia.

that were anchored in close proximity to British ships wherever they were anchored. Britain countered with a similar order and over 1,000 U.S. ships were seized by French and British privateers. The U.S. government countered this by issuing an embargo that denied all foreign ships permission to enter any U.S. ports. With that order, the constant flow of goods into New Orleans abruptly stopped. In March 1809, Congress repealed the embargo act but replaced it with another act that still banned trade with France and Britain.

New Orleans and Barataria

By the fall of 1809, Jean and Pierre were finally together and well established in New Orleans. While Pierre remained in town handling the business end of the operation, Jean organized a pirate base on Grand Terre and Grand Isle, two small islands at the entrance of Barataria Bay. That was an ideal location for a pirate base as it was on the Gulf of Mexico and accessible to New Orleans through several backwater routes. It was also a perfect place to conduct smuggling operations. Ships entering Barataria Bay could easily evade the customs inspectors whose inspection station at the port of Balize was located 30 miles to the east at the mouth of the Mississippi. Today, nothing remains of Balize as a hurricane destroyed the island in 1860.

1809

Jean Laffite establishes a pirate base in Barataria Bay

Jean quickly amassed a privateer fleet that operated legitimately, sailing under letters of marque issued by Simón Bolívar at Cartagena. Unfortunately for pirate enthusiasts who envision a huge pirate fortress at Barataria as portrayed in films and publications, there was no large fortress. Laffite's pirate

Figure 24: *New Orleans and Barataria 1814 Map*

empire in Barataria was simply a series of small wooden huts constructed on dozens of small islands spread throughout the bay. There were between 300 and 400 pirates inhabiting Barataria Bay at any one time with perhaps a dozen vessels of varying sizes. That number depended upon who was in port unloading captured cargo and who was at sea hunting for vessels to capture. There was only one real entrance to the bay from the Gulf of Mexico, a channel that ran between Grand Isle and Grand Terre that was less than 10 feet deep. That prevented any large warship from attacking or even entering the bay. Even so, the entrance was protected by three small naval guns, mounted in a wooden gun emplacement. Unimpressive by any standards, those guns would not have been effective against a large force, but did serve to discourage single vessels from attack.

Because of the need for vessels with a shallow draft, Laffite's privateers usually used schooners or hermaphrodite brigs. Both are similar in structure with two masts, only schooners are fully rigged with fore-and-aft sails while the hermaphrodite brig is square-rigged on the fore-mast and rigged with fore-and-aft sails on the main mast. They also used feluccas for smaller operations and to transport their captured goods to New Orleans.

The felucca is a small vessel of Mediterranean design that has one mast rigged with a large triangular lateen sail.

Laffite's house was on Grand Isle. Across the channel was Grand Terre where the fortifications were located. Grand Terre had the only deep-water port and served as the center of all privateer activity in Barataria. Grand Terre was where the privateers conducted business, socialized, and routinely held auctions to sell their captured merchandise. Local merchants flocked to

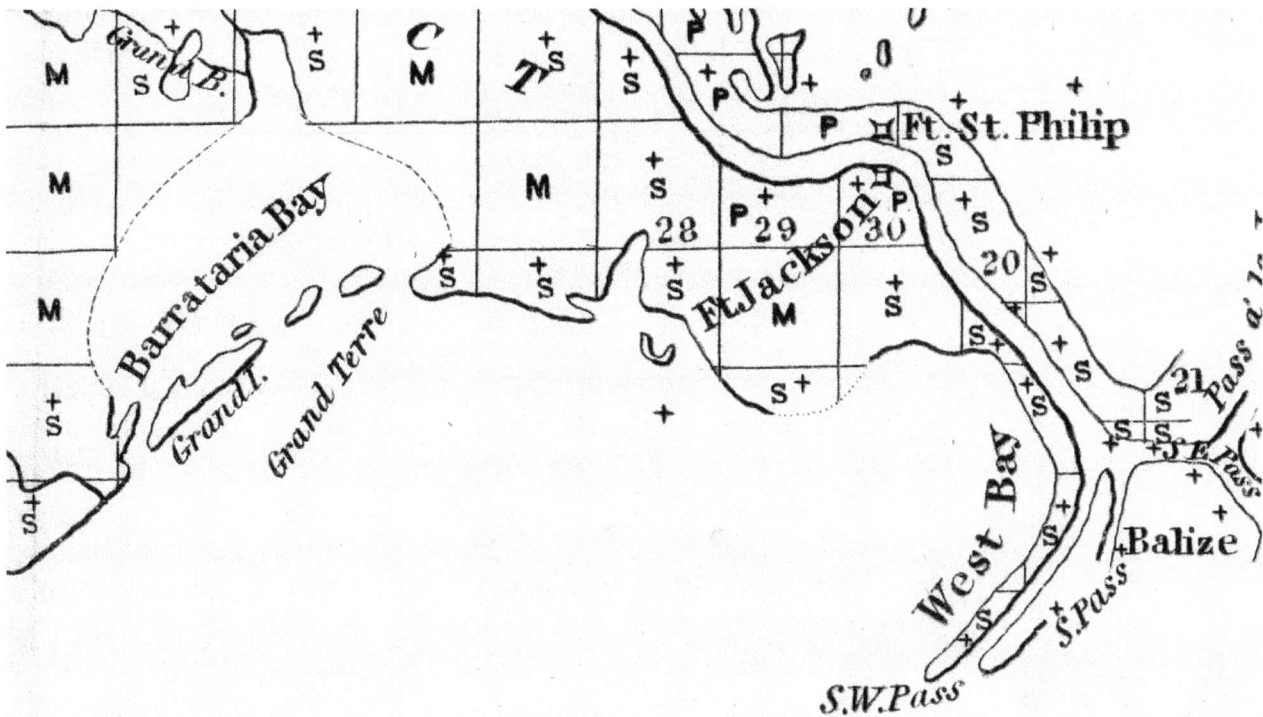

Figure 25: *Grand Isle and Grand Terre 1853 Map*

these auctions to buy goods for prices much lower than their actual value. Laffite's men were eager to get back to sea and capture more merchandise, so making rapid sales for low prices was far more profitable than spending time bargaining for better deals. In addition to the local merchants, many well respected towns folk and plantation owners from all over Louisiana attended too.

But Grand Terre wasn't the only place where the privateers sold their goods. Quite often, they would load their goods onto small boats and sail to the north end of Barataria Bay where the blue water seemed to dissolve into a dense green jungle of hundreds of twisting bayous and thousands of tiny islands. Those islands provided the ideal cover for secret trade negotiations. The privateers would set up private auctions to selected merchants on the high ground of many of those islands. Even today, locals living in the area around the town of Lafitte, Louisiana and throughout Barataria Bay are still finding pieces of jewelry and other trade items on these islands that were apparently dropped or left behind for some unknown reason.

Sometimes, goods were smuggled from Barataria directly to New Orleans for sale by Jean's brother, Pierre. There were four water routes they could take to reach New Orleans without passing any military posts or customs inspectors. The first was from the Grand Bayou through Lake Salvador to the Mississippi River near Carrollton. The second was through Wilkinson Bayou to the northeast. The third was through Big Bayou and the fourth was through Little Bayou and Bayou Rigolets. Even though the officials condemned the illegal sale of these goods, the people loved it. New Orleans had become a pirate friendly town. The fact that there were no formal charges made against the Laffites during that time proves this.

While Pierre seldom traveled to Barataria, Jean often traveled to New Orleans. While in town, he stayed with his brother and was treated like a celebrity by most of the town's citizens. Well-dressed with impeccable style and grace, Jean was a handsome and dashing figure. He attended many high society social events and could often be seen in ballrooms and at gaming tables.

It wasn't long before other privateers and pirates joined Laffite at Barataria. Most of them were former French privateers from Guadalupe which was seized in February of 1810 by the British. As the British closed the French privateer base of Guadalupe, the privateers had to look for another base of operations and New Orleans was the most likely choice. Among those French privateers who left Guadalupe and resettled in Barataria was the master gunner Dominique You. Another one was Louis-Michel Aury, who arrived in Barataria in May of 1810. He served with Jean Laffite for several years, then went on to other things which eventually led him to Amelia Island in Florida. We shall read more about him in a later chapter.

In February 1810, three French privateers entered the Mississippi and stopped at the customs office on Balize. They claimed to be legitimate privateers with letters of marque and asked permission to enter port and sell their goods. But upon close inspection, the officials found that they were carrying blank letters of marque which could be filled out to legitimize whatever vessel they took. Apparently, the pirates had neglected to fill in the appropriate information. The pirates were arrested, but local opinion in support of pirate operations was so strong that they were eventually released.

During his time at Barataria, Jean Laffite seldom went to sea, if at all. He was too busy running his operations and overseeing the sales of his goods at Grand Terre and in town. If he did go to sea, there is no historic record of it. But his privateers certainly did. They mainly operated in the Caribbean, taking British prizes near Jamaica and Spanish prizes all around Cuba, but they also took prizes off the west coast of Florida in the Gulf of Mexico.

Cedar Key on the west coast of Florida was most likely used by many of Laffite's privateers, possibly Renato Beluche and Louis-Michel Aury, to ambush passing Spanish vessels.

One of the first to join Laffite was Renato Beluche. Born in New Orleans in 1780, Renato's father was a wig maker and a successful and rich smuggler. When Renato was a young boy, his father actually owned the plantation at Chalmette where the battle of New Orleans took place. In the early 1800s, Renato lived in a large house in the city and was already an experienced ship's master and smuggler by the time Jean Laffite arrived in 1809. Beluche was among Laffite's most successful privateers. Local legend contends that many of Laffite's privateers, specifically Renato Beluche, often used Cedar Key, Florida to ambush Spanish vessels just as the privateers of William Augustus Bowles had done a decade earlier.

Cedar Key was the perfect spot from which a privateer could ambush passing vessels. The two most important Spanish ports in West Florida were San Marcos and Pensacola. Vessels traveling to and from those ports often stopped at the freshwater spring near the mouth of the Anclote River to take on fresh water. As you may recall from the previous chapter, the Anclote River is located on the west coast of Florida about 40 miles north of the entrance to Tampa Bay. Vessels would have to set a course fairly close to the west coast in order to stop there. That course would bring them near the islands around Cedar Key.

One of Laffite's French privateer captains was cruising along the coast of the Florida panhandle in 1811 aboard his sloop *La Franchise*. The sloop was spotted by a U.S. gunboat which gave chase. The *La Franchise* made a desperate attempt to make it back to Barataria but the gunboat was too fast. As *La Franchise* passed Pensacola Beach it was about to be overtaken.

The captain realized that they couldn't get away, so he ran the sloop aground, set fire to it, and he and his crew simply disappeared into the low-lying shrubbery of the barrier island.

Life and business were good for the Laffites until 1814. Britain and the United States had been at war since July 1812 and Britain's two recent attempts at an invasion in New York and Maryland had both failed. That left Louisiana as the next most likely target. You may recall the British strategic plan they had with William Augustus Bowles which was to encourage large Native American forces to drive up from Florida and down from Canada and secure the territory between the United States and the Mississippi River, which would eventually become a British protectorate. That plan failed, but now it seemed as if it could be resurrected. The new plan was for a large British naval and military force to first take New Orleans, and then slowly drive northwards, securing the territory along the Mississippi River and isolating the United States to the east coast. In the summer of 1814, the British Admiralty began preparations to send approximately 15,000 British troops aboard 60 vessels to attack New Orleans.

Intelligence reports from British agents in Pensacola identified Barataria as a key location from which to launch an attack. Those same reports identified Jean Laffite as the man in charge of the privateers who occupied that bay. At first, the British attempted to make a deal with Laffite. They sent Captain Lockyer aboard his brig, the *HMS Sophie*, to offer Laffite a deal. The British warship arrived at Barataria on September 3, 1814 and Laffite was offered gold, a commission in the British Navy, and land grants for his assistance. Laffite stalled the anxious British naval officer and asked for a week or so to make a final decision. After the *HMS Sophie* departed, Laffite took the written British offer to Governor Claiborne in New Orleans and informed the governor of the British plans. Laffite assured the governor of his loyalty to the United States.

FEBRUARY 6, 1810

British naval forces seized the French colony of Guadalupe

Meanwhile, Commodore Patterson of the United States Navy decided not to take any chances on Laffite's loyalty. He was aware of the British offer and didn't trust the pirates to follow through with any of their promises. Patterson thought that the written offer Laffite produced may have even been a forgery. Additionally, he was worried that the British would land at Barataria regardless of Laffite's decision. Before the British could establish a foothold in Louisiana, Patterson ordered an attack against Barataria. His force was comprised of about 70 regular soldiers aboard the USS *Carolina*, the USS *Sea Horse*, and 6 gunboats.

The early morning of September 15, 1814 seemed like any other morning to the privateers of Barataria. There were perhaps 400 people spread throughout the bay. The shores of Grand Terre were busy with merchants from town who were carefully looking over a fresh haul of merchandise. Nearby, a cargo of wine and rum from a freshly captured Spanish brig was being unloaded. Other vessels were being repaired or were taking on supplies and preparing to sail. Suddenly, at about 8:30 in the morning, the small American fleet was sighted. At first, confusion and uncertainty quickly spread throughout the privateers and merchants ashore and those aboard the vessels. But as the naval force approached, that uncertainty quickly turned into panic. Laffite was still in New Orleans and there was no one to organize a defense. Some rushed towards the nearest small boat and prepared to escape into the bay. Those aboard larger vessels prepared to escape to the open sea. Others decided to fight and began to form a defensive line with the few vessels that were seaworthy.

By 10:00 a.m., Patterson's forces were near enough to open fire. They could see the panic and disorder on the beaches and aboard the vessels. Some privateers raised a white flag and surrendered. Delighted to see this, Patterson also raised a white flag, indicating that he would accept their surrender. But as Patterson's vessels neared, some of the privateers and merchants began to flee and a mass exodus suddenly ensued. Some fled into the swamp on foot and others attempted to sail away in small boats into the many hidden and uncharted bayous. One of those vessels in particular was a felucca carrying Dominique You. That vessel prompted one of the gunboats to open fire, as they didn't want to see any pirates get away. Other gunboats also opened fire. By noon it was all over and Patterson's men were in complete control of Barataria. They confiscated 27 vessels which included a Spanish prize brig that was in the process of being unloaded and eight schooners, two of them with their privateer crews still on board. In total, 80 privateers had been captured including Dominique You.

In addition to the vessels and men, the American force confiscated a fortune in stolen goods. Aboard one vessel alone, they recovered a cargo of cloves, coffee, 17,200 cigars, casks of sherry and wine, 769 gallons of brandy, raisins, candles, 483 pounds of chocolate, and $1,641 in the currency of the day in coins and bank notes. It is estimated that the value of the goods captured by Paterson equaled $200,000 in terms of their value in 1814. All of it went to help fund the defense of New Orleans against the British.

General Andrew Jackson and his meager military force arrived in early December 1814. He had a daunting task, to defend the city against 15,000 of the best trained troops in Europe. To make matters worse, Jackson didn't have enough supplies to fight the British. He was in desperate need of muskets, musket flints, and gunpowder and Jackson was told that there just weren't any of those available in New Orleans. However, that wasn't

SEPTEMBER 15, 1814

U.S. naval forces under command of Commodore Daniel Patterson destroy Barataria

The Battle of New Orleans was a series of engagements which began with the Battle of Lake Borgne on December 14, 1814, and ended with the main battle at Chalmette on January 8, 1815.

First opened in 1788, Pierre Maspero's is still in operation and is considered to be one of the finest bars in New Orleans.

DECEMBER 24, 1814

Jackson begins to construct his defenses behind the Rodriguez Canal along the Chalmette Road near a plantation that was once owned by Renato Beluche's father.

Artillery pieces were often referred to by the weight of the projectile they fired. For example, a 6-pounder gun fires a projectile that weighs six pounds and an 18-pounder fires a projectile that weighs eighteen pounds.

completely true. Jean Laffite had plenty of both. Jackson had recently hired a translator so he could speak with the French-speaking residents of the city. That interpreter just happened to be a friend of Jean and Pierre Laffite and a meeting was soon arranged between General Jackson and the Laffite brothers. That meeting took place in Pierre Maspero's, a local tavern on Chartres Street in the center of the city.

Jackson and the Laffites quietly sat in an upstairs room and planned the defense of New Orleans. Jackson was in no position to be too particular about the help he received, so an agreement was quickly reached. All of Laffite's men would be released and a pardon would be offered for anyone who fought on the side of the United States. In return, Laffite would bring 7,500 musket flints, 500 muskets, and a storeroom of gunpowder to the front line. These were supplies that Laffite had hidden somewhere in the city. Laffite was good to his word. The supplies were delivered to Jackson at his defensive position behind the Rodrigues Canal along the Chalmette Road. Additionally, 36 pirates joined his ranks including Jean Laffite, Dominique You, and Renato Beluche. Both Dominique You and Renato Beluche were given command of two of the gun batteries.

Beluche must have played a vital role in the planning and preparation of Jackson's defenses. Beluche's father once owned the plantation that Jackson was now using as his headquarters. As a child, Beluche played on the very grounds that Jackson was preparing for battle. His in-depth knowledge of the terrain and the surrounding bayous would have been invaluable.

On the cold, crisp morning of December 28, 1814, Jackson was still preparing his position with only a few of his guns in place. Suddenly, the British appeared and advanced in two columns. They were supported by fire from their field artillery and by Congreve rockets, which were screeching weapons designed to create panic among inexperienced troops. When Jackson ordered his guns to return fire, a terrific cascade of metal tore into the British ranks. Later, one of the British soldiers called it "As destructive a fire of artillery as I have ever witnessed." One battery in particular concentrated on the British artillery and swiftly knocked out two six-pounders and sent the crew of two nine-pounders fleeing. That accurate fire came from the battery commanded by Dominique You. Afterwards, the British fell back to their staging area about four miles away.

Four days later, the British returned with larger guns and dozens of Congreve rocket launchers. An artillery battle between the British and U.S. forces ensued. Fire from the British guns was far more intense and their rockets pierced the tree line and fell indiscriminately onto waiting American troops. Once again, it was Dominique You who ended the engagement. His accurate fire destroyed several of the larger British guns. Then, in a spectacular demonstration of marksmanship, You fired a single shot towards the spot where rockets were coming from. Seconds later, a series of huge explosions occurred and hundreds of rockets shot out of the smoke in every direction. Apparently, You had scored a direct hit on the ammunition supply for the rockets.

Jackson was of course victorious and the British finally left Louisiana. Laffite and his men were pardoned; however, the Laffites were no longer able to use Barataria as a privateer base. The Laffite brothers had to find another source of income. Meanwhile, the revolutionary forces of Simón Bolívar in Cartagena and those in Mexico continued to grow stronger. Simón Bolívar was still issuing letters of marque to all who applied, and privateers and piracy were running rampant throughout the entire Caribbean and Gulf of Mexico. Many of Laffite's privateers chose to stay in their profession after the operations on Barataria were closed. They joined other privateers who were sailing in support of one or another of the many revolutionary forces fighting against Spain. The most famous of those men was Louis-Michel Aury. In 1815, he joined Simón Bolívar's forces at Cartagena and later joined the forces of the Mexican War for Independence. By 1717, Aury had established a modest pirate base on Galveston Island and was taking vessels throughout the Gulf of Mexico.

Galveston: The Laffites and the Mexican War for Independence

At this time in history, many historians and authors who have written accounts about Jean Laffite simply state that Jean decided to return to piracy by 1717 and to move in on Aury's operations at Galveston. In this simple version, Laffite sailed to Galveston, kicked Aury out of power, and took control of the base, declaring himself the new "Bos" of the pirates of Galveston. But that is not at all accurate or even truthful. The real truth is far more complex.

Let's go back to February 1815. The British threat was gone and everyone in New Orleans began to turn their attention towards the Mexican War for Independence. The entire Caribbean and Mexico had been in turmoil since 1808 with revolts against Spanish rule. In Mexico, the Mexican War for Independence had transformed into a highly political and chaotic revolution at all levels, with many powerful politicians and high-ranking military officials switching sides between support for Spain and support for the revolution. In the midst of the confusion, the anti-Spanish faction had temporarily taken control of the government. To counter this, Spain stepped

Congreve Rockets were developed by Sir William Congreve in 1804 and were commonly used by British forces in the War of 1812 and the Napoleonic Wars.

1810–1821

The Mexican War for Independence was a series of individual and uncoordinated insurgencies against Spain.

Jose Manuel de Herrera was an official in the government of the Mexican War for Independence. In 1815, he was sent to New Orleans to negotiate with officials from the United States government to obtain arms and ammunition. He played a vital role in the movement until it was successful in 1821 when he was appointed Minister of Foreign and Internal Relations of the new Mexican Republic.

up their military operations and the newly formed Mexican government was in trouble and needed support and funding. One of the original leaders of the movement, Bernardo Gutiérrez, had already traveled to Washington D.C. and had requested official assistance from the United States government. Another leader of the movement was Jose Manuel de Herrera, who was sent to New Orleans as the emissary from the newly formed Mexican government to the United States.

Several years earlier, while France and Spain were at war, Napoleon had sent several spies to Louisiana with orders to encourage and support the Mexican War for Independence. Napoleon thought that a successful Mexican revolution would pull valuable Spanish resources away from the battlefields in Europe. In early 1815, Napoleon was preparing France for war again and a weaker Spain was vital to his success. French agents in New Orleans were more active than ever in garnishing support for the Mexicans. Recruiting efforts made by the Mexican leaders and French spies were very successful and hundreds of armed Americans were waiting in Louisiana to join the Mexican struggle.

On top of all that, the United States government became secretly involved. Not wanting to spark a conflict with Spain, the Madison administration couldn't officially take sides or give any aid to the Mexicans. But secretly, the Madison administration wanted the Mexican War for Independence to succeed. The United States government had laid claim to much of Texas, believing that it was included in the Louisiana Purchase, but Spain refused to grant them access. Madison hoped that he would have more success with a new Mexican government. Even though the United States would not officially provide aid, their support came in the form of "looking the other way" while others did. Government officials did little to prevent privateers from smuggling their stolen goods into New Orleans and selling them for profits that benefited the revolution.

By the summer of 1815, New Orleans had become the center of all covert activity and international espionage as far as the Mexican War for Independence was concerned. In addition to profits from privateer goods, there were also profits from the sale of arms and gunrunning into Mexico. Plus, there was the opportunity for great profit in the sale of information on both sides. In the middle of all this intrigue were the Laffite brothers. Spying and gunrunning make for big profits and the Laffites were eager to join in on the action and take their cut.

The Laffite brothers had long been associated with all the pro-Mexican agents in Louisiana and continued to discuss business arrangements with them. However, Spanish agents approached both Laffite brothers independently in the summer of 1815 and offered to engage them as spies for Spain. Double agents can make lots of money, so the Laffites accepted. While planning their support for the Mexicans with a group of men known as the "Associates," the Laffites were secretly passing the details of those plans on to the Spanish.

By 1816, the Mexican War for Independence was falling apart. Pro-Spanish factions had regained control of the government and the rebels were on the run. The ousted Mexican government's emissary in New Orleans, Jose Herrera, planned to get things going again by establishing a secret base somewhere along the Gulf coast, although the exact location had not yet been selected. He planned to use privateers to attack Spanish shipping and raise money for the revolution. He also planned to use this base as a staging area from which he could launch a full-scale attack against the Spanish in Mexico. Herrera chose the well-known privateer Louis-Michel Aury as his leader of privateers and sent him a message. Aury had been serving as a privateer for Bolívar in Cartagena, but things weren't going so well for him. The Spanish had recently taken Cartagena and Aury needed to find a new place to continue his fight against the Spanish, so he accepted Herrera's offer.

In the summer of 1816, while en route to join Herrera somewhere along the Mexican coast, Aury was very successful taking Spanish shipping in the Gulf of Mexico. With a fleet of 15 vessels, he had captured 10 Spanish prize ships and was holding 150 captives aboard his vessels. As Aury approached the Mexican coast, he spotted what seemed to be the perfect spot for a pirate

DECEMBER 1815

Spanish forces capture Cartagena after a five-month siege

Figure 26: *Gulf of Mexico 1791 Map*

base. It was a large, flat island with soft white sand, wide beaches, and a gentle surf. There were few trees on the island. The ground beyond the beach was blanketed with a lush, low vegetation. Hidden far away from the main shipping routes, the island was only inhabited by a few transient fishermen. The nearby Karankawa Indians were friendly and could provide food. But what really made the island the perfect choice for a pirate base was the deep-water channel running behind the seaward side that was ideal for the construction of a port. That island was named San Luis and the bay behind the island was named Galveztown after Benardo Gálvez y Madrid, Count of Gálvez. Even though Galveston didn't become the official name of the island until the late 1820s, I shall refer to it as Galveston to avoid confusion.

1816

Louis-Michel Aury establishes a base on Galveston

Aury sailed his fleet toward what he hoped would be the location of his new base and landed on August 5, 1816. Unfortunately, he wasn't familiar with the offshore sandbars and ran most of his vessels aground. To make matters worse for Aury, most of his crew mutinied a month later and shot Aury in both hands and in the chest. Severely injured, Aury laid quietly in his tent, most of his men gone, the beach in a shamble, and several beached abandoned vessels swaying nearby in the surf. Then, on September 9, 1816, Herrera arrived bringing supplies and news of strong military support. Aury quickly made a full recovery and Herrera and Aury began planning their invasion of Mexico and building a privateer base. Within a month, Galveston was operational.

Aury's base on Galveston wasn't quite the massive fortification that some have come to believe. In actuality, it was very unimpressive. By the end of Aury's command, there were only about 120 small structures scattered around the island, some made of planking with sailcloth roofs and others built like native huts with thatched roofs. Those buildings were used as living quarters, administrative offices, warehouses, dry goods stores, and one was even a coffee house. The only fortifications were two unimpressive earthwork gun emplacements mounting a total of 6 naval guns near the entrance of the bay. Even though the base was uninspiring, dozens of privateer vessels soon flocked to Galveston to join Aury, most of them were his old friends from either Barataria or Guadalupe.

To say that Aury's base on Galveston was merely a pirate base would be totally untrue. In reality, it was the headquarters of the Mexican revolutionary government. Herrera and other leadership within the movement were always present. In accordance with Herrera's plan, the base served two functions. The privateers sailing from Galveston generated the funding the revolution needed while the island itself served as a staging area where an army could be formed to eventually invade Mexico and revitalize the stalled revolution.

Louis-Michel Aury was officially named the governor of Galveston and made the senior admiral of the Mexican Navy. As governor, Aury himself was authorized to issue letters of marque, which he printed on his own printing press. Aury was also appointed to the office of customs inspector and he constructed a customs house where he inspected all goods that were brought to the island by his privateers. Aury also established an Admiralty Court and appointed Herrera's legal secretary, John Peter Rousselin, to serve as judge. Rousselin tried all privateers who had committed offenses at sea that violated their captains' established rules of conduct.

By October 1816, Aury's privateers were taking vessels all over the Gulf of Mexico. This excluded any American vessels. Aury was aware that his best customers were the citizens of New Orleans and American merchants would hesitate to buy any stolen goods taken from American vessels. Additionally, the Mexican fight for independence was still hoping for official support from the United States, so the taking of any American vessels was strictly forbidden. Spanish vessels were the prime targets of Aury's privateers. It was part of the strategic plan to support the revolution by taking money, goods, and resources away from the enemy while funding the revolution. As the Spanish grew weaker the Mexicans grew stronger. With Aury's privateering success, the next step in the process to regain control of the Mexican government was to launch a military invasion of Mexico itself. Galveston was used as the staging area for that campaign.

MAY 1816

Martín Francisco Javier Mina is appointed commanding general of the Mexican army

The overall command of the military forces and consequently the Mexican army, was given to General Francisco Javier Mina. Back in May of 1816, Mina had met with representatives of the Mexican government in Baltimore and was offered the position of commanding general. Exceptionally well qualified, Mina was brought in from Spain to save the Mexican War for Independence. He had been the leader of the guerilla fighting against the Spanish government under French control for decades. As an outsider, he wouldn't be subject to all the infighting and backstabbing that had plagued the movement in recent years. Mina landed on Galveston in late November of 1816 with a shipment of 2,500 muskets, 9 artillery pieces, 50 barrels of gunpowder, and 50 soldiers. However, General Mina and Louis-Michel Aury didn't get along right from the first. Each one claimed to be in charge. This conflict set the stage for Aury's failure and the eventual destruction of Aury's base.

I must take a moment to remind the reader of the technical differences between pirates and privateers. A privateer is a legal pirate. Privateers carry letters of marque from legitimate governments giving them legal permission to attack shipping from enemy nations named in those letters. Those who sailed without letters of marque were definitely categorized as pirates. But what about those who sailed with letters of marque issued by illegitimate or unrecognized governments? Were they pirates or privateers? That is far more difficult to classify. If a specific nation chose to recognize the letters of marque as legitimate, then they were privateers, but only for that nation. For all other nations, they were considered pirates. That made their status of either a pirate or privateer a highly political matter. In 1816, the Mexican Republic was not recognized as a legitimate nation by anyone, so technically, Aury's men were pirates. However, to the New Orleans merchants who purchased Aury's stolen goods, the Mexican letters of marque that Aury himself printed and issued gave his pirates the pretense of legitimacy and that was good enough for the merchants.

New Orleans was still a pirate friendly town where goods could be easily smuggled into the city. Barataria was no longer operational as a base, but all the old smuggling routes were still very active and the merchants in town were still eager to receive stolen goods. Aury didn't keep any of his captured goods on Galveston, he took them to Barataria and arranged for their sale in New Orleans. Occasionally he would sell to his old friend, Pierre Laffite. One particular vessel arrived at Barataria with $320,000 in goods in terms of their value in the early 19th century. After making the sale, $200,000 went to the captain and crew, while the remaining $120,000 was deposited in the Bank of Louisiana to pay the Mexican revolution's backers in town.

The main financial supporters of the Mexican War for Independence in New Orleans were a group of American merchants known as the Associates. Some of them were just in it for the profits while others were idealistic and wanted to see the new republic succeed. Eventually, these differences began to pull the group apart. They would meet regularly with Mexican agents in New Orleans acting on behalf of Aury, Herrera, and General Mina. One of the members of that group was Pierre Laffite. But unknown to Aury, the Mexican agents, General Mina, or any of the other Associates, both Pierre and Jean Laffite were actually Spanish spies. While Pierre was pretending to support the Mexican revolt, he was in reality gathering intelligence for the Spanish in New Orleans. Pierre even had a code name, Number 13.

However, Jean was not in New Orleans during that time. Throughout most of the year of 1817 Jean was on a secret mission to Arkansas, gathering intelligence for the Spanish on American expansion. During that mission, he actually obtained copies of the secret maps of Pike's Peak that Zebulon Pike produced after his expedition ten years earlier. When Jean returned to New Orleans in January 1817, he once again became heavily involved with his brother gathering intelligence on the Mexican revolution for Spain. By coincidence, just a month after Jean returned, the biggest opportunity to gather intelligence walked right into Jean's hands. General Mina traveled to New Orleans to personally meet with the Associates and to discuss his plan for the invasion of Mexico. Both Pierre and Jean were invited to attend.

Jean must have been absolutely shocked. There he was, sitting at a meeting with the senior officer of the Mexican army hearing all the details of his invasion plans. For a spy, opportunities to hear extremely valuable information directly from the source don't happen very often. But before he made his report to the Spanish, Jean wanted to gather first-hand information on Galveston itself. By traveling to Galveston, Jean would be able to see the whole picture and determine the exact condition and strength of the Mexican revolution. So as not to raise suspicion, Jean offered to deliver a shipment of wine and other supplies directly to Aury's base. General Mina left New Orleans and sailed for Galveston on March 1, 1817. A little over two weeks later, on March 16, 1817, Jean sailed for Galveston aboard the schooner *Devorador*.

MARCH 20, 1817

Jean Laffite arrived at Galveston

Four days later, as the *Devorador* cut through the deep blue waters of the Gulf, the white sails of another vessel topped the horizon. That vessel was the *Bellona*, and it was heading directly towards Laffite's schooner. As the two vessels neared each other, the captain of the *Bellona* sent an urgent signal to Laffite's schooner to stop and to come alongside. A short time later, the two vessels rested beside each other, securely tied together. The highly agitated captain boarded Laffite's vessel and rushed to meet him. He identified himself as one of Aury's privateers and told Laffite that he had just left Galveston in a state of total chaos. He told Jean the astounding story of the conflict between Aury and Mina. A few days earlier, as General Mina's ship returned, Aury ordered his men to fire upon him. Apparently, Aury didn't want to allow Mina to come ashore and resume their older disagreements. The captain of the *Bellona* who was telling the story to Laffite was actually the one who Aury had ordered to fire the shots. Immediately afterwards, that captain quickly sailed the *Bellona* out of port, deciding to escape the bedlam while he could. The shocking news must have stunned Jean. Anxious to learn the truth about the condition on Galveston, Jean ordered the *Devorador* to proceed cautiously. As his schooner reached Galveston Bay, Jean could see six vessels anchored in the harbor. Not wanting to become involved in a tense and potentially dangerous situation ashore, Jean decided to drop anchor near the shore and simply wait and watch.

After laying at anchor for several days, Jean decided to chance a landing. He quickly learned that the situation on Galveston was actually far worse than expected. He was told of the events that led up to the *Bellona's* rapid departure. While Mina was away in New Orleans, Aury planned on abandoning Galveston, leaving Mina's men ashore. When one of his captains, a man named Perry, voiced his objections, Aury had him arrested. That action caused an immediate split of all of his forces. Those loyal to Aury and those who supported Perry and Mina stood face to face with weapons drawn and guns pointed. One wrong move would have triggered a volley of fire and would have quickly erupted into a massive battle. But Aury chose to back off and both sides returned to their huts. A few days later, surprised at Mina's early return, Aury ordered the men aboard the *Bellona* to fire on Mina's ship. After firing, the *Bellona* quickly sailed out of port, eventually reaching Jean at sea. During that time, Mina remained on board while tension among everyone ashore remained extremely elevated.

Eventually, General Mina came ashore and announced that he was going to take his army of about 200 men and invade Mexico at Tampico. He told the privateers that he planned on sailing in just a few days. Perry and his followers agreed to sail with him. Aury must have seen this as an opportunity to finally rid himself of the tenuous relationship with Mina, so he agreed to accompany them until they reached Tampico and then to separate and establish a new base on the island of Matagorda, just south of Galveston. The entire force left Galveston on April 7, 1817, taking everything of value with them, including all the artillery. Only about 30–40 men remained behind. They were a mix of officials in the Mexican government and a few of Aury's men who apparently had decided that the time was right to leave Aury's command. But before Aury left, he burned most of the buildings, clearly indicating his plans never to return. Within a few hours, Aury and Mina's fleet disappeared over the horizon. A single vessel remained anchored in the lonely bay. It was Jean Laffite's vessel, the *Devorador*.

APRIL 7, 1817

Louis-Michel Aury abandons Galveston

Just one month earlier, Galveston Bay was crowded with vessels and the beach was littered with tons of boxes of cargo that were either being unloaded or loaded aboard the privateer vessels. Activity was everywhere, with hundreds of privateers busily preparing their vessels to sail or just socializing with their friends. All of that was now gone. Smoke from over 100 recently burned structures choked the air. The 30–40 bewildered men who were left behind must have been wondering what they would do next. The beach was totally empty, except for a scattering of broken or useless debris that was cast aside in their haste to leave.

JULY 15, 1806–JULY 1, 1807

The Pike Expedition was a military party led by Zebulon Pike that explored the Great Plains and the Rocky Mountains.

It is impossible to know exactly what Jean Laffite was thinking as he walked along the beach. Most likely his first reactions were that of shock and amazement at what he had just witnessed. But that certainly didn't last long. Laffite was a man of action and was quick to see opportunities. For a pirate in 1817, Galveston was the opportunity of a lifetime. Aury was correct, it was the perfect spot for a pirate base. Laffite moved quickly.

Campeche (Galveston) Jean

Laffite's New Base

Laffite named his new base Campeche. The island actually wasn't named Galveston until about 1830. His first step was to get the support of those men who Aury had left behind, including several of the less important leaders of the Mexican revolution. One of them was John Peter Rousselin, who had been Aury's Admiralty Court Justice, Herrera's secretary, and a loyal supporter of Mexican independence. Since support for the Mexican independent government gave a certain legitimacy to Aury's operations, Laffite decided to continue the pretense. To accomplish this, Laffite needed Rousselin to help him establish a relationship with the independent Mexican government and to reestablish a base on Campeche (Galveston). Laffite offered to make Rousselin the new customs inspector with a percentage in pay of the value of everything he inspected. That did the trick. With Rousselin now loyal to Laffite, they arranged for Barthelemy Lafon, Herrera's assistant, to come to Campeche (Galveston). He arrived On April 15, 1817, and administered the oath of allegiance to Mexico to everyone. That is, everyone except Jean Laffite. Now, his vessels would all have letters of marque and they would sail under the flag of the independent Mexican government.

Next, Laffite made arrangements with the Association back in New Orleans to completely dissolve their support for Aury and shift their support to him. With legitimacy from the Mexicans and a means of financial support, the last thing to do was to actually build a pirate base. Laffite had learned a lot about building and organizing a base while at Barataria; he learned what worked and what didn't. This new base was to be different. Laffite planned on correcting all the errors he had made before. The pirates of Barataria had not been well organized and they had no real plan for their defense. That's why it was so easily taken by Patterson in 1814. To correct this, Laffite appointed a military commander who would lead a defense force that would not only protect the island from attack but would enforce order and discipline among his pirates. This defense force would be paid voluntarily from all of the pirate's shares.

As far as his allegiance to an independent Mexico goes, Laffite personally believed that the movement had already failed. He never had any intention of sending on any profits to them. He planned on keeping everything for himself. Furthermore, unknown to his Mexican allies, Laffite was still a spy for Spain. All of the plans the Mexican leaders made with Laffite were eventually forwarded on to the Spanish agents in New Orleans. While pretending to be working directly with the Mexican officials, Jean Laffite was actually working for the Spanish.

Apparently, the Spanish weren't concerned about Jean Laffite taking their vessels throughout the Gulf of Mexico and the Caribbean as long as he continued to feed them inside information on the revolution. They probably felt that their vessels would have been taken by pirates anyway and that the information that the Laffites provided was worth the loss. But on top of pretending to support the revolution while reporting to Spain, Jean Laffite was actually working for himself. He viewed Campeche (Galveston) as his own personal island, not part of Mexico or any other nation. Laffite envisioned it as becoming the ideal pirate realm with himself as the ruler.

As soon as Campeche (Galveston) began to take shape, Jean returned to New Orleans to confer with his brother and to give his report to the Spanish. He arrived back in New Orleans on April 22, 1817. Shortly afterwards, Jean pitched his plan to his Spanish contact. His brother Pierre would sail to Matagorda to check on Aury's progress while the Spanish government would furnish him with several well-armed vessels which he would take to Galveston. Those Spanish vessels would appear to be recently captured prizes to the people on shore. Once in port, the Spanish would arrest the Mexican officials and all their allies. Since most of the privateers were Laffite's old comrades from Barataria, he believed that they would join him without any resistance. The part of the plan that he didn't tell the Spanish was that after securing Galveston, he intended to keep all the Spanish vessels and give them to his privateers. Then, he intended to build Campeche (Galveston) up as his pirate base.

Meanwhile, on May 4, 1817, as Jean discussed his plan with the Spanish in New Orleans, Louis-Michel Aury returned to Galveston to reclaim the island. He was shocked at the rapid changes. Aury argued that he was still governor, but the Mexican leaders that Laffite had placed in charge refuse to recognize his authority. After two weeks of sulking around the island, Aury finally left for Matagorda. But when he arrived, Aury ran most of his vessels aground on the sandbars that surrounded the island, just as he had done when he first arrived at Galveston the previous year. Acting on the information that the Laffites had provided, a Spanish frigate accompanied by two gunboats arrived at Matagorda on June 12, 1817, and attacked Aury's grounded vessels. Having inflicted sufficient damage, the Spanish left two days later. Pierre Laffite's timing was perfect. He arrived at Matagorda as the Spanish vessels were sailing away.

In late June 1817, Aury had managed to repair some of his vessels and sailed back to Galveston along with Pierre. Once on shore, Aury again announced that he was in charge of the island. Pierre played along with Aury. He really wasn't concerned because he knew that Jean and the Spanish vessels were due any day. But by mid-July 1817, before Jean or any Spanish vessels arrived, Aury finally lost confidence in the success of the Mexican revolution and decided to once again return to Cartagena and sail under the colors of Simón Bolívar, this time as part of Bolívar's new plan to invade Florida at Amelia Island. On July 31, 1817, Louis-Michel Aury sailed away from Galveston, never to return.

With Pierre now in charge on Campeche (Galveston), pirate vessels began to arrive. But food was rapidly running out. Pierre returned to New Orleans to get more supplies. Meanwhile, Jean's plan to use the Spanish to attack Galveston fell through. Apparently, the Spanish didn't completely trust the Laffites and were stringing them along. However, that went both ways. The Laffites knew where they stood with the Spanish and were playing it cool. But Campeche (Galveston) was still the Laffite's main concern. With news of his men starving, Jean sailed from New Orleans aboard the *Carmalita* on September 1, 1817. When he arrived a week later, he was shocked to see that Campeche (Galveston) was totally abandoned. He would have to start all over again. With only about 40 men, Jean began to rebuild the base.

But it wasn't the pirates who had the biggest impact on the growth of Campeche (Galveston), it was the merchants in Baltimore and Philadelphia. There were laws that prevented American merchants from selling guns and ammunition, as well as other similar supplies, to privateers in U.S. territory, but those laws didn't prevent them from selling to pirates on foreign soil. They stood to make a fortune if Galveston became the pirate base that Jean Laffite envisioned. One ship arrived from Baltimore carrying sugar, liquor, beef, pork, muskets, cannon shot, and enough building supplies to build a real base. Those supplies included tar, turpentine, pitch, resin, varnish, and 11,500 feet of lumber. But the involvement of the merchants from Baltimore and Philadelphia went far beyond selling supplies. They became involved with the Argentine War of Independence.

By 1816, Argentina and Chile finally decided to join together and fight against Spanish rule. With Buenos Aires as their headquarters, agents from their revolutionary government began issuing letters of marque directly to any privateer who asked. In Cartagena, letters of marque from Simón Bolívar were increasingly becoming harder to get. Bolívar had experienced the problems caused by issuing letters to untrustworthy pirates and had become far more selective as to who received his letters of marque. Not so for the struggling government in Buenos Aires, whose officials became the prime source of easily obtained letters of marque for pirates. They even placed agents in Baltimore and Philadelphia who could issue letters directly. Fourteen commissions were issued from those cities in the fall of 1817 alone and those privateers returned with 42 captured Spanish prizes.

SEPTEMBER 1817

Jean Laffite begins to build a pirate base on Galveston

JULY 9, 1816–APRIL 5, 1818

Argentine War of Independence from Spain

111

Meanwhile, Laffite was making progress building a base on Campeche (Galveston). In early January 1818, one of Laffite's privateers captured a Spanish vessel and took it to Campeche (Galveston) with the captain and crew still on board. The captain was Manuel Gonzalez and his captured vessel arrived on January 6, 1818. Captain Gonzalez later described Laffite's base as having several small huts, some used for housing and others used for storage. There was only one real structure on the island, the house of Jean Laffite. It was a good-looking frame house built on a small rise, with two stories and a piazza in the center. The house had a large storeroom filled with muskets and ammunition. There were only two women on the island, a Mademoiselle Victorie, who ran a small cabaret, and a slave woman.

Construction of Laffite's privateer base on Campeche (Galveston) continued and more and more privateers as well as pirates began using it as their main port. Some had letters of marque and some probably didn't. That didn't seem to make a great deal of difference to Jean. As Laffite's operations grew, he really didn't command them or even have a large group of his own privateers. Although many of them were his friends and would sail aboard his vessel as his crew when he needed to travel back and forth to New Orleans, most of them were independent and used the port at Campeche (Galveston) as the most convenient place to sell their goods. Laffite would buy their goods and either take them to Barataria to be smuggled into New Orleans and sold by Pierre just as in the old days, or Jean would sell them to the occasional merchant who sailed to Campeche (Galveston) to get some bargains. In the Spring of 1818, one of those bargain hunting merchants from Nacogdoches stopped at Campeche (Galveston) and later wrote a description of the port. It had several gun emplacements at the entrance to the bay, a wooden dock with structures on it, a coffee house, several shops, and far more women. Some of them were actually wives of the privateers.

During that time, the last vestiges of the old Mexican War for Independence faded away and their supporters finally lost all hope. Even Herrera turned himself in to the Spanish authorities. But the loss of the Mexican officials didn't bother Laffite. He continued to print and issue his own letters of marque in the name of the non-existent Mexican government. This technically turned all of his privateers into true pirates. But that didn't seem to matter. One of the captured vessels brought to Galveston was the *Campeche*, which contained a cargo worth $360,000 in the value of the day. But the profits from those valuable cargos didn't seem to find their way into the pockets of the Laffites. Most of the money they earned went right back into building supplies for their base. In April 1818, Pierre had accumulated so much debt that he had to sell his house.

With the threat of the Mexican revolution finally gone, the Spanish were no longer paying the Laffites as spies. But just as one threat faded, another one arose. A group of about 300 French exiles who had supported Napoleon and had fled France after his arrest, decided to carve out a French settlement for themselves in Mexico. Called "Napoleonic Exiles" or "Bonapartist Refugees," they weren't refugees in the classical way we normally think of refugees. They were former senior military officers, wealthy businessmen, tradesmen, and soldiers along with all their families. They were also well financed and had connections with those back in France who had remained loyal to Napoleon.

The first group of those French refugees had actually arrived on Galveston back in January of 1818. Laffite provided assistance by giving them food, but both camps remained separate on opposite ends of the island. Relations between the French and the pirates were tense but peaceful. By the late spring of 1818, part of their group had traveled about 30 miles up the Trinity River from Galveston and established a base of 28 buildings manned by about 120 soldiers and their families. They also had fortified their settlement on Galveston by constructing a 90-foot square fort. Their existence threatened Spain, as they intended to settle in Texas and eventually take over the entire colony and make it a French colony loyal to Napoleon. They even planned on rescuing Napoleon from his prison on Saint Helena Island and bringing him to Texas to lead their new colony.

The Laffites saw this as another opportunity to get money as spies for the Spanish government. In early June 1818, Pierre took over operations on Galveston and Jean sailed to New Orleans to once again become a Spanish spy. Shortly after his arrival, Jean devised a plan with the Spanish to get rid of the French. However, there was another interested party, the United States.

The border between Louisiana and Texas had been hotly debated between the governments of Spain and the United States since the Louisiana Purchase. The Monroe administration secretly hoped that the Mexicans would take control and negotiate a new boundary. Now with Spain firmly in control of Mexico, the Monroe administration decided to use direct force to take possession of the territory that they felt was within the Louisiana Purchase. That included most of Texas. In May of 1818, General Jackson, with a large force of U.S. troops, invaded Florida and captured Pensacola under the pretense that he was in pursuit of hostile Seminoles as part of the First Seminole War. Then, the United States government turned their attention towards Galveston. The presence of the French interlopers concerned the Madison administration even more than the presence of Jean Laffite.

1816–1818 First Seminole War

Andrew Jackson invaded Florida in March of 1818 and destroyed Mikasuki villages near Tallahassee, seized the Spanish post at San Marcos, and attacked and occupied Pensacola.

But before the Spanish or the Americans had the chance to deal with the French settlers, their colony collapsed. The lack of supplies and the harsh conditions caused most of the settlers to simply give up. When Jean Laffite returned to Galveston on July 20, 1818, he found the French camp had been abandoned. A month later, on August 24, 1818, a special agent for the United States, George Graham, reached Galveston to make deals and free the island for occupation by the United States. Upon arrival, he observed that Laffite had securely grounded a large well-armed brig on a sandbar parallel to the shore. It mounted 18 guns and had total command of the entire island, as the gun deck sat above the flat and treeless terrain. As soon as Graham arrived, he set up a small camp near Laffite's port and sent a message to Laffite requesting a meeting.

George Graham (1770–1830) was a friend of President Madison and had served as the chief clerk for both the Secretary of State and the Secretary of the War Department.

Laffite answered Graham's request for a meeting immediately. Everyone who personally met Jean Laffite usually commented on his amazing graciousness and good manners. This occasion was no exception. The meeting was exceptionally cordial, and Jean pointed out that he had helped save New Orleans in the recent war with Britain. He also pointed out that his privateers never took an American vessel. He did admit to smuggling but proposed that everyone did it. He also expressed his desire not to go against the wishes of the United States.

The two of them came to a quick agreement. Laffite had three months to gather his possessions and make arrangements to leave, then he would relocate to any territory not belonging to the United States. In the meantime, he would cease all smuggling of goods into U.S. territory. But stopping Laffite from taking Spanish ships was another matter. As part of the U.S. government's campaign against the Spanish, Graham recommended that Laffite obtain new letters of marque from the Buenos Aires agents in Philadelphia and continue attacking Spanish shipping. Graham even wrote him a letter of recommendation. But before Laffite could gather his possessions, a hurricane hit Galveston in early September 1818 and destroyed everything except the grounded brig, one small sloop, and six structures. One of those surviving structures was Laffite's house.

Meanwhile, the Spanish sent a letter to the United States Congress in October 1818 stating that they intended on attacking Galveston and getting rid of everyone. Acting quickly on the news, Pierre Laffite traveled to Washington and offered to spy for the United States against the Spanish. Immediately afterwards, Pierre, the perfect double agent, went to the Spanish ambassador in Washington and told him everything the Americans were planning. He also told the Spanish that Graham had asked Jean to team up with Louis-Michel Aury to attack Spanish shipping, which was probably a total lie. Since his activities were becoming compromised, he requested that his secret code name be changed from number 13 to number 19. Pierre's new plan was for Spain to fund a new pirate base on Galveston, allow hundreds of privateers to flock there, then attack and destroy them all at once. Also, he would make sure that the Americans were implicated in the scandal.

All this espionage and intrigue came to a halt when the United States and Spanish governments finally came to an agreement on the precise location of the border between Texas and Louisiana. Conflict between the two nations faded and American support for other revolutionaries such as those in Buenos Aires evaporated. Meanwhile, Jean was still rebuilding Galveston. With no threat from the Americans, he decided to go back to business as usual. He was actually joined by his old friend Dominique You for a while.

During that time, Jean went into the slave trading business. The United States passed laws restricting the importation of any slaves into the country. That meant that all new arrivals from Africa had to be smuggled into Louisiana from Texas or by landing them ashore directly from the Gulf along the many miles of isolated shoreline. Slavery was still legal in the Spanish empire, so slave ships in the Caribbean were common. The operation was simple. Take a Spanish slave ship, then take the slaves to the Texas border where they could be smuggled into the United States through Nacogdoches. One of his partners in the Texas operation was the famous Jim Bowie. According to local tradition, there is also an isolated spot along the beach near modern-day Gulfport, Mississippi that was used by Jean Laffite to smuggle slaves into the country.

Galveston was as strong and prosperous as ever by August 1819. Both Jean and Pierre were now living on the island. Just as a decade earlier, many of the vessels that Laffite's pirates captured were taken in the Gulf of Mexico, traveling to either Pensacola or San Marcos. As before, it is likely that some of his pirates used hiding places all along the west coast of Florida to wait in ambush for passing vessels. Local legend contends that Cedar Key, Tampa Bay, and even the islands and inlets south of Tampa were commonly used. However, I have not been able to find any direct evidence of this.

All of the pirates from Galveston were using letters of marque signed by Laffite himself, which were not recognized as legal by anyone. In the summer of 1819, the United States decided to really begin cracking down on piracy. On August 29, 1819, one of Laffite's pirates, Captain Jean Defarges, took a prize 100 miles off Key West. It was the *Filomena* and was loaded with large quantities of food as well as $3,000 in gold coins in the value of the day. Two days later, the pirates were boarded by two U.S. revenue schooners, the *Alabama* and the *Louisiana*. Upon search, they found letters of marque signed by Laffite as well as older letters signed by Aury. This was the first

FEBRUARY 22, 1819

Adams-Onís Treaty between the United States and Spain ceded Florida to the U.S. and defined the boundary between the U.S. and Spanish territory. It became effective on February 22, 1821.

Jean Defarges (sometimes spelled Le Farges) and most of his crew were hanged for piracy between January and June of 1820.

instance where the American authorities had absolute proof that Laffite was involved in piracy. Shortly afterwards, all pirates entering port were closely inspected and arrested if their letters of marque weren't in order.

The standard version of the U.S. reaction which appears in many publications is that the forces of the United States blasted Laffite off the island. Nothing could be further from the truth. By the end of 1819, Jean Laffite realized that he wouldn't be able to remain on Galveston for long. On January 3, 1820, he sent a message to Commodore Patterson requesting a letter of safe conduct to allow him to relocate. Commodore Patterson was the senior naval officer in Louisiana and was the commander who destroyed Barataria back in 1814. Patterson took the message to the governor and other officials. They all agreed to meet Laffite's request. On January 24, 1820, the letter of safe conduct was sent to Laffite. The only requirements were that he destroy his entire base on Galveston, stop taking prizes illegally, and settle outside of United States jurisdiction.

The USS *Enterprise* was launched in 1799. It was originally built as a 12-gun schooner but was rebuilt as a 14-gun brig in 1812.

As Laffite made preparations to leave, he caught sight of a U.S. warship entering port. It was the USS *Enterprise* under command of Lt. Lawrence Kearny. He was sent to ensure that Laffite was indeed leaving. Not wanting to miss an opportunity to entertain a guest for dinner, Jean Laffite invited Lt. Kearny aboard his brig. As usual, the meeting was very cordial and everyone enjoyed themselves. The next day, Laffite took the Lieutenant ashore and showed him their progress. The beach was filled with supplies ready to be loaded and there were signs everywhere that the base was being dismantled.

Only three pirate vessels remained in the bay. The largest was the 18-gun brig which had been grounded and used as protection. Laffite refloated it and renamed it the *General Victoria,* then took it as his personal vessel. He also had two schooners, the *Minerva* and the *Blanque.* Jean kept his word about leaving Galveston and on May 7, 1820, he burned all the remaining buildings and sailed away. However, he didn't keep his word about not taking prizes illegally. On May 22, 1820, he took a Spanish felucca near Tampico and dropped the crew off on the Mexican coast a little to the north. Then, he headed right back to Galveston. After transferring all the stolen cargo to the *Minerva,* the captured vessel was burned, and the *Minerva* sailed away to rejoin Laffite.

Mujeres, Cuba, and Cartagena

In July 1820, while cruising along the Campeche coast, Laffite's three vessels spotted a larger and well-armed Spanish merchant vessel. In an uncharacteristic move, Laffite decided to ask his crew if they wanted to risk the attack. During the discussion, Laffite himself revealed that their letters of marque were illegitimate. Not wanting to be hung as pirates, thirty-one of his crew voted to leave. The three vessels sailed east to the island of Mujeres, which is a tiny island about 5 miles west of Cancún, Mexico. It was decided that Laffite and his followers would keep the two schooners and the others would be allowed to leave in the brig. As they weren't going to continue as pirates, all the guns onboard the brig were transferred over to the schooners. Once at sea, Laffite's well-armed schooners overtook the unarmed brig and ordered that the crew destroy all the rigging and throw the masts overboard. Apparently, Laffite wasn't very happy about their decision to leave, which he viewed as mutiny.

Cancún was settled by Spanish colonists in the early 18th century.

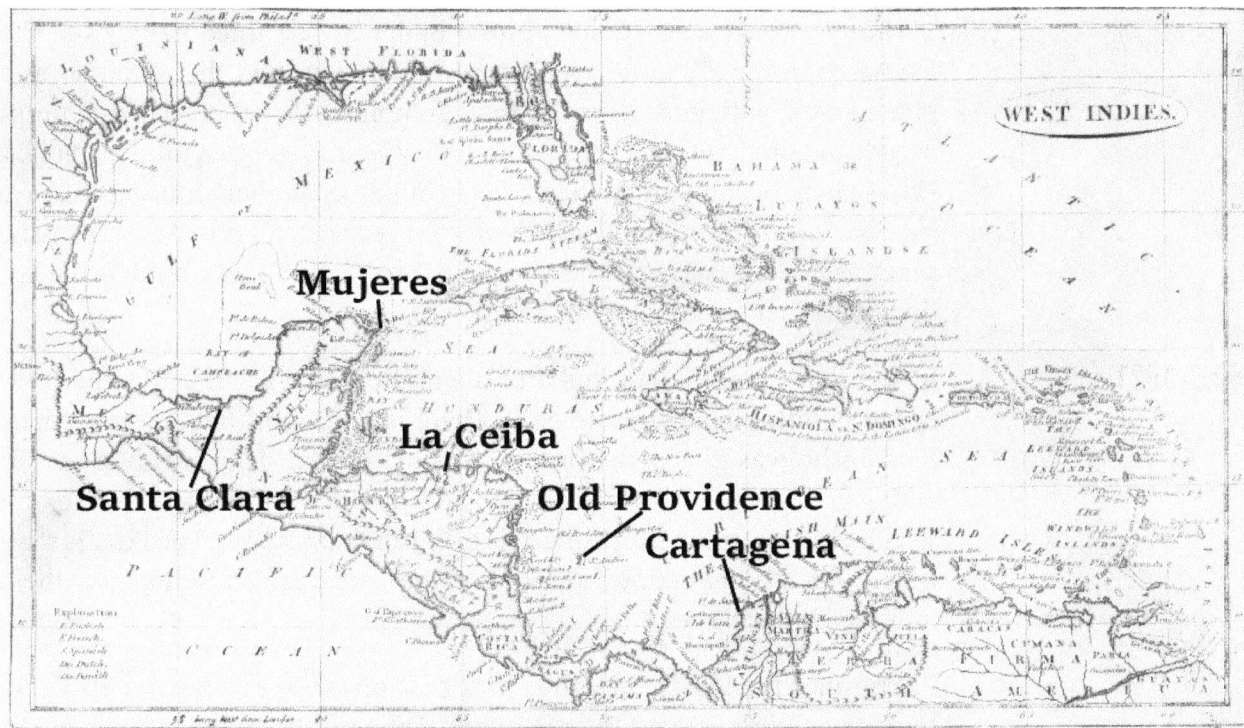

Figure 27: *Caribbean 1799 Map*

Jean Laffite decided to establish a new pirate base on Mujeres. Why not? The small village of Cancún was nearby and he could sell his merchandise there. Plus, it was far enough away from the Spanish authorities that he felt the island would be a safe place for a base. He sent word to Pierre who had returned to New Orleans and then sailed back to Galveston to rendezvous with his brother. Pierre sailed into Galveston aboard the *Two Friends* in August 1820 where Jean and Pierre were reunited. They agreed to wrap up all their affairs in New Orleans and shift their operations permanently to Mujeres. However, to legitimize their piracy, they would need letters of marque. They agreed that the new legal source of their letters would be from their old comrade, Louis-Michel Aury.

As mentioned earlier, Aury left Galveston back in July 1817 and sailed to Amelia Island, Florida. As you will read in the next chapter, that operation ended in disaster for Aury. He had fallen out of favor with just about everyone except Jose Cortes de Madariaga, who was among the leaders of the combined insurgencies of Buenos Aires and Chile. Aury was able to obtain letters of marque from Buenos Aires and he relocated to Old Providence Island. Upon arrival, Aury proclaimed himself governor. The Laffites believed that they might be able to obtain some legitimate letters of marque from him.

MARCH 1818

Louis-Michel Aury arrives on Old Providence

Meanwhile, Pierre sold everything in New Orleans and sailed to Charleston, South Carolina, arriving on November 15, 1820. He was unknown there and had a better chance of doing business. While in Charleston, he purchased a schooner named the *Nancy Eleanor* under the assumed name of Pierre Francisco. After the purchase, Pierre illegally outfitted the schooner as a pirate vessel, with guns, ammunition, grappling hooks, extra bilge pumps needed to pump out water in a battle, and other things often used by pirates. He also bought enough supplies to begin building another pirate base. Pierre left Charleston aboard the *Nancy Eleanor* on February 21, 1821. Accompanying him was his new lady friend, a woman named Lucia. That would be the last time Pierre set foot on American soil.

MARCH 1821

Jean and Pierre reunite on Mujeres

Jean and Pierre were reunited on Mujeres in March of 1821. Together they began building a new pirate base. There is no doubt that it was actually Pierre who was giving the orders. By June of 1821, Jean was back at sea taking prizes while Pierre was making deals with local farmers to help him sell his stolen cargo. Meanwhile, it is believed that Jean returned to his slave trading operations, taking slave ships in the Caribbean and smuggling them into the United States.

Pierre's luck finally ran out. The pirate base on Mujeres never grew to anything resembling Galveston or Barataria. When Jean was out at sea, perhaps only a half dozen pirates remained in port. On October 30, 1821, a small group of Spanish soldiers from a nearby port found the base on Mujeres and attacked it. They captured Pierre, his lady friend Lucia, and 3 of Pierre's pirates. The next day, an armed pirate vessel sailed into port and realized that the Spanish had captured Pierre and were occupying their base. The pirates let loose a broadside of grapeshot and the Spanish fled. Pierre may have either been wounded in the battle with the Spanish, wounded by the broadside from the pirates, or was already ill from an unidentified disease he had recently contracted. Either way, he was on the verge of death. The recent fighting left most of their huts in ruins, so the pirates carried Pierre onboard and abandon their tiny base. Pierre died at sea on November 9, 1821. He was buried in the churchyard of a Franciscan Mission in Santa Clara, a small city near the Mexican coast between Vera Cruz and Tabasco.

Unaware of the fate of his beloved brother, Jean was cruising with his two schooners in the Caribbean taking prizes. Somewhere between the Cayman Islands and Jamaica, Jean Laffite engaged a vessel that fought back. In a surprise broadside that caught the pirates unprepared, several pirates were killed. Eventually the vessel was captured by Laffite's men and they took it to Santa Cruz del Sur on the south coast of Cuba to ransom it to the owners. But another surprise awaited Laffite's men. As they approached shore, they were met by two warships that opened fire. One of Laffite's schooners sank and his casualties were heavy. With his second schooner badly damaged, Laffite lowered his longboat into the water and escaped to the shore. A short time afterwards, he was captured and taken to jail at Porto Principe, which was also named Villa del Principe and was later renamed Camagüey in the 20th century. Jean became very ill and was transported to an infirmary, where security wasn't quite so tight. While there, a few unidentified friends helped him escape on February 12, 1822.

By March 1822, Jean was once again the captain of a pirate schooner sailing out of Cayo Romano, a well-known pirate base on the northeastern coast of Cuba. The circumstances are unclear as to why this group of pirates made Jean a captain so quickly. Perhaps his reputation as the most famous pirate in the world had something to do with it. On April 11, 1822, while sailing with three other pirate vessels, Jean Laffite did something he had never done before. He took an American prize. It was the *Jay,* and the vessel was taken off Gibara, which is on the northeastern coast of Cuba. In retaliation, the USS *Alligator* and the USS *Grampus* sought out Laffite's schooner and attacked on May 1, 1822. With a shallower draft, Laffite's schooner could sail over shoals that the others could not, and he quickly escaped his pursuers.

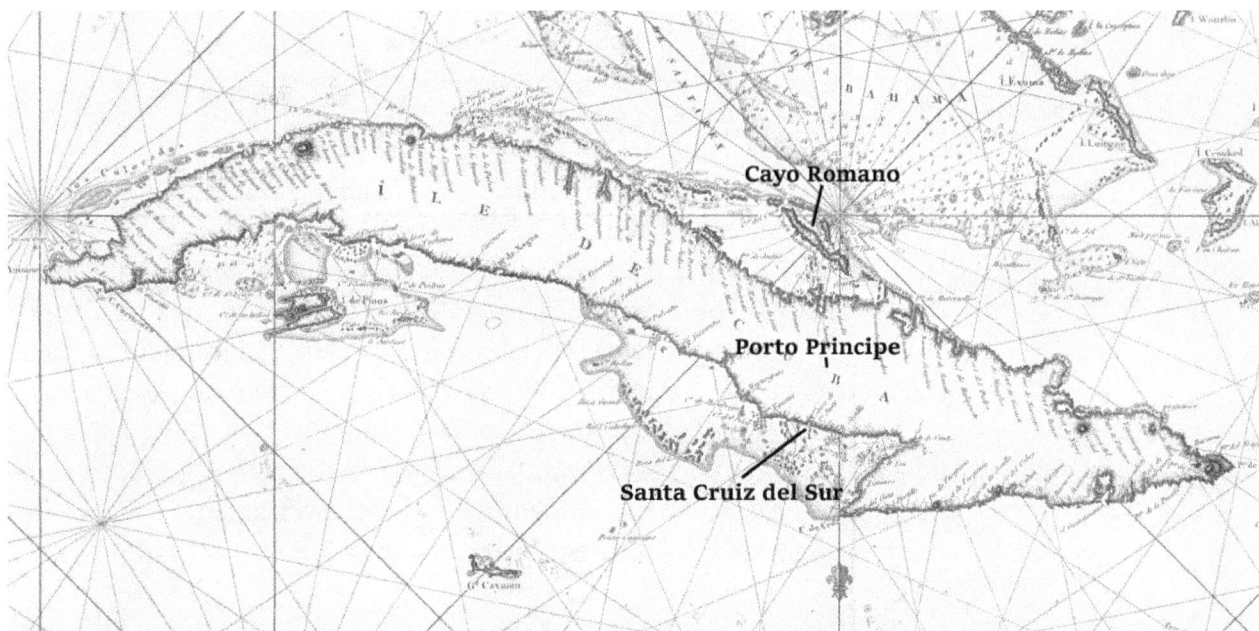

Figure 28: *Cuba 1801 Map*

SEPTEMBER 1822

Jean Laffite joins the
Colombian Navy

After the harrowing experience of being attacked by U.S. warships, Laffite came to the conclusion that the Cuban waters were just too dangerous. Plus, he really needed to become legitimate once again.

Simón Bolívar, leader of the combined revolutions in Central and South America against Spain, had successfully waged war at sea through the unlimited use of privateers. As early as 1808, he issued letters of marque to just about anyone who wanted to attack Spanish shipping. Bolívar's need to use privateers changed in 1821 as victory was in sight. Bolívar felt the need to bolster his international status as a legitimate head of state. He decided to stop issuing any letters of marque and to simply hire privateers to serve in his navy. There is a significant distinction between being a privateer sailing under letters of marque and serving in a nation's navy. Privateers with letters of marque generally operated under their own initiative. They could go anywhere they wanted and attack any vessels listed in their letters. There was no national control of their actions by any government official. However, naval officers, whether they were former privateers or not, were tightly controlled under the authority of a head of state. The captains who sailed for Simón Bolívar after 1821 operated completely under the orders of Bolívar alone. They didn't have letters of marque because they didn't need them. They were official members of his Colombian Navy, which was centralized in Cartagena.

Some of Jean Laffite's old friends from Barataria had already joined the Colombian Navy, the most notable being Renato Beluche who was now officially an admiral. Jean Laffite joined them and took command of the schooner *General Santander* on September 28, 1822. His vessel was a recently captured Spanish prize armed with one 18-pounder and a 4-pounder swivel gun and had a crew of 30 men. For the first time in his life, Jean went to sea as the captain without the need of any letters of marque. He was the captain of a legitimate naval vessel.

Shortly afterwards, there was a report that Laffite captured an American brig named the *Sampson* off the Dry Tortugas in the Florida Straits. The date was August 19, 1822, and according to the captain of the *Sampson*, the pirate captain identified himself as the famous Jean Laffite and mistreated his crew horribly. It is doubtful that this person was in fact the real Jean Laffite. There is no record of Jean Laffite treating any of his captives harshly. Additionally, he had just accepted a commission in the Colombian Navy. If he did in fact take an American vessel, he would have been tried and hanged as a pirate. Most likely, this was some unidentified pirate who wanted to impress his captives and hide his true identity at the same time.

Further evidence that this was not the real Jean Laffite can be seen in the events that occurred only two months later. On November 26, 1822, while sailing 50 miles south of Grand Cayman, Jean Laffite encountered an American schooner named the *Columbus Ross*. The two vessels came alongside each other and exchanged greetings. In characteristic form, Laffite was charming, elegant, and friendly. He gave the captain of the *Columbus Ross* a few supplies they needed and then safely escorted the schooner through pirate waters. Would Laffite attack an American vessel one month and be friendly and helpful to another two months later?

NOVEMBER 9, 1821

Pierre Laffite dies

Just like Pierre, Jean's luck was about to run out. While cruising off the Gulf of Honduras, 40 miles west of La Ceiba, Laffite's schooner spotted two unidentified vessels, a brigantine and a schooner. He gave the order to pursue. After 17 hours of chase, night had fallen, and the sea was totally dark. The only lights were the dim lights coming from the bridge of the enemy brigantine. The other vessel had sailed out of sight. Laffite's schooner finally came within gunshot range of the brigantine and opened fire. The brigantine returned fire, then raised several lanterns high into the rigging. That was a signal for the other schooner to return. Within moments, Laffite realized that he had been caught in a trap. The other schooner suddenly appeared out of the darkness and opened fire. As the solid cannon shot crashed through the planking, a large splinter flew through the air and pierced Laffite's chest. Outgunned and with the captain severely wounded and his second in command killed, the first mate took command of the *General Santander* and ordered the schooner to break contact with the other vessels. The next day, on February 4, 1823, the most famous pirate of the century quietly died, and Jean Laffite was buried at sea. With his passing, the last era of piracy in the Caribbean came to an end.

One question remains. Did Jean Laffite visit the coasts of Florida? As with Blackbeard and William Kidd, stories about Jean Laffite burying treasure along the east and west coast of Florida abound. The majority of those stories identify Amelia Island as a place that Laffite often used. After all, Laffite's fellow pirate, Louis-Michel Aury, was definitely there. Why not Laffite? So far, there has been no direct evidence that Jean Laffite ever visited Amelia Island. I would have to say that just like Captain Kidd and Blackbeard, Jean Laffite did not visit Amelia Island.

FEBRUARY 4, 1823

Jean Laffite dies

As far as the west coast of Florida is concerned, Cedar Key was likely used by Laffite's privateers such as Renato Beluche, but it is doubtful that Laffite himself was ever there after 1809. It is impossible to say if Laffite used Cedar Key or any other Florida location to ambush Spanish prizes during his earlier years before the establishment of Barataria. He primarily stayed in the Caribbean. However, if the Jean Lafitte who sailed into New Orleans in 1804 is the same Jean we are speaking of, it proves that he was taking prizes in the Gulf of Mexico too. Since few records of his actions during those years have been found, it is possible that Jean Laffite occasionally used Cedar Key or other Florida locations in the Gulf to ambush prizes. After 1809, his actions are very well documented. While at Barataria and Galveston, he seldom if ever actually took a prize vessel. He was too busy handling the smuggling end of the operation or overseeing his base and managing his privateers. In the final years, all reports of Laffite's piracy are restricted to the Caribbean.

What about his treasure? Could Jean Laffite have buried any treasure along the Florida coasts? Although it has been proven that some pirates and privateers did bury treasure from time to time, I don't believe that Laffite did. There was no reason for him to do so. Before 1809 he was a French privateer sailing for his nation out of Guadalupe. While at Barataria and Galveston, Laffite had a well-established system of smuggling and financing which rendered the need to bury treasure totally unnecessary. Afterwards, he was in Mexico and Cuba operating as a pirate and then in the Caribbean sailing for the Colombian Navy with no opportunity to bury any treasure anywhere in Florida.

Chapter 22
Amelia Island's Pirate, Louis-Michel Aury

Tranquil, charming, and quaint, Amelia Island has amazing beaches, a wide assortment of hotels and Bed & Breakfasts, wonderful restaurants, and a long and rich history. One of the barrier islands, it is located on the east coast of Florida just below the Georgia border. It is the ideal spot for tourists who appreciate the finer things and are looking for an out-of-the-way spot. There are several museums and the remnants of an early 19th century Spanish fort. There is also a large Civil War period fort, Fort Clinch, which was constructed in 1847 and offers an exciting look back into American history.

Locals throughout the island and even several visitor websites claim that the island has long been associated with pirates. They name William Kidd, Blackbeard, and Jean Laffite as the primary pirates who used the island to ambush passing vessels and to bury their treasure. These legends were mostly products of the tourism industry and are both fascinating and somewhat appealing. Previously in this book, I have dealt with the facts surrounding each of these pirates and it is my opinion that they never came ashore on Amelia Island. However, there is one pirate who did. His name was Louis-Michel Aury and there is absolutely no question that he occupied Amelia Island and used it as a base.

The Timucuans had been living on the island for hundreds of years when Spanish Franciscans arrived in 1573 and established the mission of Santa Maria de Sena. The relationship between the Franciscans and the Timucuan people was a friendly one. However, the mission was abandoned in 1702 after the English from Charles Town began attacking Spanish settlements in preparation for their assault on St. Augustine.

In 1716 along Florida's east coast, there was a lot of pirate activity around the sunken treasure fleet. Those who came too late to find anything generally turned to piracy to make a living out of necessity. The main trade route leaving the Caribbean was still through the Florida Straits, between South Florida and the Bahamas. Some pirates may have waited in hidden inlets all along the east Florida coast to prey on ships. Local tradition tells us that pirates often used Amelia Island as an ambush spot and occasionally buried treasure there. This is possible, as it is strategically located near the main trade route and the natural deep harbor hidden from view is actually ideal for that type of operation. Additionally, it was far enough away from the Spanish authorities at St. Augustine and in the early 18th century, there were only a few inhabitants living on the island. However, so far, no contemporary reports or accounts have been uncovered to confirm that 18th century pirates used Amelia Island as an ambush spot. As for the treasure, lots of small caches of treasure have been found on the Island but it is impossible to tell if they were buried by pirates or by plantation owners hiding their valuables during times of war. The island was invaded several times by substantial military forces between 1702 and 1865.

FEBRUARY 10, 1763

Treaty of Paris ended the Seven Years' War and ceded Florida to Britain

Figure 29: *Close-up of an 1860 map Showing Amelia Island*

Ownership of the Island was hotly contested between the British and Spanish until 1763 when all of Florida was ceded to Great Britain. Almost immediately afterwards, Amelia Island began to be settled by the British. By 1770 there were plantations and a village at the site of the future town of Fernandina. During the Revolutionary War, the island was invaded by continental forces from Georgia in 1776 and many of the homes were destroyed; however, Spanish control of Florida returned in 1783 as a result of the end of that war. The British inhabitants, along with hundreds of British loyalists who fled from Georgia, used Amelia Island as an embarkation point to leave for other British colonies. In the process, they destroyed all the buildings and within a few months, only one family remained on the island.

Amelia Island was named for Princess Amelia, daughter of King George II of Britain.

The Spanish slowly returned and by 1795 there was a small Spanish garrison on the Island which continued to grow in size. In 1801, a small wooden fort with three guns was built to protect the rapidly developing town of Fernandina. Feeling the need to bolster up their defense of this strategic spot, the Spanish built the much larger concrete fort of San Carlos in 1816 that mounted four long 16-pounders, five 4-pounders, and one 6-pound carronade. The fort was also flanked by two blockhouses which mounted 4-pounder guns.

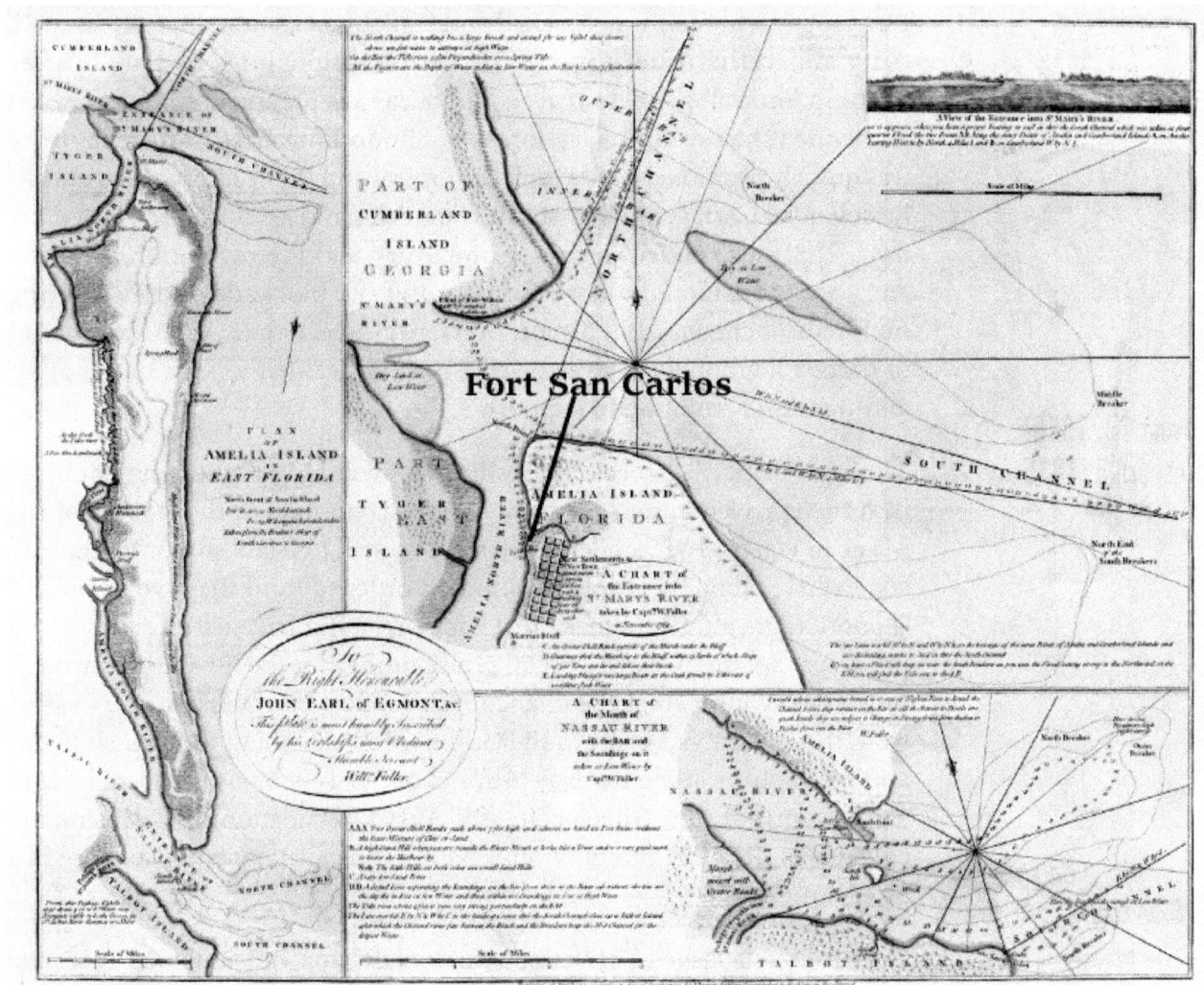

Figure 30: *Location of Fort San Carlos*

Louis-Michel Aury's activities in New Orleans and Galveston were covered in the previous chapter. As for his early years, Aury was born in Paris around 1788. Bitterly anti-Spanish, he joined the French Navy in 1802 and served in the Caribbean. At some point, he transferred over to a French privateer vessel and by 1804 was sailing from San Domingue. In the midst of the Haitian slave revolt, Aury helped evacuate many of the refugees from Haiti to New Orleans and elsewhere. It is possible that he may have met Pierre Laffite at that time. Afterwards, he sailed with French privateers out of Guadalupe where he probably was introduced to Jean Laffite. On February 6, 1810, as part of the war against France, a large British naval force seized the French colony of Guadalupe and Aury lost his vessel. Now, in addition to his hatred of the Spanish, he hated the British too.

By 1810 he was the captain of his own vessel, the *William*, and joined Jean Laffite at Barataria. Just like most of Laffite's privateers, Aury sailed under letters of marque from the Colombian government of Simón Bolívar. In 1811, Aury's schooner was attacked and sunk by a U.S. warship. Afterwards, Aury added the Americans to the list of nationalities he hated. After Commodore Patterson destroyed Barataria in September of 1814, Aury left and joined the revolutionary forces of Simón Bolívar at Cartagena. Aury was quickly promoted to the rank of admiral and in the last months of 1815, he defended Cartagena against a full Spanish naval assault which lasted 105 days. That siege ended in disaster for Aury and the revolution. Even so, Aury was considered a hero for running the blockade and evacuating hundreds of citizens. But Simón Bolívar refused to pay Aury the amount of money he thought he was owed, so Aury decided to seek out another anti-Spanish revolution to support.

AUGUST 23, 1815– DECEMBER 5, 1815

General Pablo Morillo laid siege to Cartagena

Timing was perfect for Aury. As he was wondering where he could go next, he received a message from Jose Herrera, one of the leaders of the Mexican War for Independence. Herrera wanted to establish a privateer base somewhere along the Texas coastline in order to fund the revolution and support a land attack. To make that happen, Herrera needed an experienced privateer captain to lead a fleet and raise money for the revolution through piracy. Aury jumped at the chance. With a fleet of 15 vessels, Aury sailed from the Colombian coast to the Gulf of Mexico, taking 10 Spanish prizes and 150 captives along the way. Aury chose Galveston as the ideal base in early August of 1816. The specifics of Aury's occupation of Galveston are covered in detail in the previous chapter, but for convenience, here is a brief summary.

Once Aury established a privateer base on Galveston, he was joined by several representatives of the Mexican government to legitimize the operation. Aury was made the governor and began issuing his own letters of marque, which he printed on his own printing press. Things were going well until November of 1816 when General Francisco Mina arrived. The Mexican revolutionaries had selected him to lead their invasion of Mexico. Aury and Mina did not hit it off. They began to argue over who was in charge right from the start. This argument escalated and eventually led to Aury ordering his men to fire on Mina's ship in March of 1817. Mina was prepared to leave for his invasion of Tampico and Aury reluctantly went with him, burning all the buildings on Galveston. That's when Jean Laffite took control of the island.

On May 4, 1817, Aury unexpectedly returned to Galveston and wanted to take control once again. At that time, Jean Laffite was in New Orleans making arrangements for supplies and his brother Pierre was running things on Galveston. Pierre and Aury argued for a while and the Mexican officials said that they now supported the Laffites. After sulking around for a while, Aury left to build a new base on Matagorda. That didn't last long, as the Spanish attacked him on June 14, 1817, and his base was destroyed. After repairing a few of his vessels, Aury returned to Galveston once again in mid-July 1817 and attempted to take control for a third time. That only lasted for two weeks. Announcing that he had lost confidence in the War for Mexican Independence, Aury decided to return to Simón Bolívar and join him in his new plan to land on Amelia Island and take Florida away from Spain.

1807-1814

The Peninsular Campaign was fought in Spain and Portugal to combat Napoleon's forces

The revolution wasn't going well for Simón Bolívar. Before 1815, Spain really couldn't do much about the revolution because the Spanish government was busy fighting Napoleon and the French occupation. But after Napoleon was defeated at the Battle of Waterloo in June 1815, Spain was free to send troops to South America. The result was the siege of Cartagena in the second half of 1815. Now, Spain was even stronger. In order to regain momentum for his revolution in South America, Bolívar needed a distraction. He chose Florida. Bolívar believed that if he could start another revolution in Florida, it might pull Spanish troops away from Venezuela. It would even be better if he could somehow involve the United States. To carry out this complex operation, he chose two men, Louis-Michel Aury and Gregor MacGregor.

Born in Scotland in 1786, Gregor MacGregor was a soldier in the British army during the Peninsular Campaign and served under the Duke of Wellington, reaching the rank of Major. In 1811, while attending a party in London, he met one of the generals from the South American revolutionary forces in Caracas. The 23-year-old MacGregor was in the habit of introducing himself as a colonel, even though he was still only a major. The revolutionary general was impressed and offered him a job, so MacGregor left for Caracas. In South America, he quickly established himself as a military leader and politician. He married Simón Bolívar's cousin and was promoted to Brigadier General. By the summer of 1816, his military and operational success had made him a celebrity among the revolutionaries. Clever, charismatic, loyal, and an accomplished military leader, MacGregor was the ideal choice to lead this risky and complex operation that Bolívar was planning. Additionally, as a Scotsman, he spoke English and could effectively communicate with the Americans.

The plan called for MacGregor to travel to the United States and recruit a small force of ex-soldiers, then invade Florida at Amelia Island. Once the island was secure, they would announce that all of Florida was an independent republic. Aury would arrive with a small fleet and provide naval support. From there, the small revolutionary force would spread southward. The Spanish would retaliate, hopefully sending troops from South America. Bolívar also hoped that the United States might even lend its weight to the operation and attempt to annex Florida. If not, MacGregor was on his own. In reality, there was never any chance of turning Florida into an independent republic but that didn't really matter. Pulling the Spanish away from Venezuela was the actual objective. The lives of the Americans that MacGregor recruited were expendable.

In the early months of 1817, MacGregor, who was accompanied by his wife, traveled throughout the southeastern United States, and recruited several hundred men primarily from Charleston and Savannah. They were former soldiers from the war of 1812 and knew battle. Accompanied by a small army of 80 men, only 55 of which had muskets, MacGregor and his wife sailed from Charleston and landed ashore near Fort San Carlos on Amelia Island on June 29, 1817. The Spanish offered no resistance. Upon seeing the approaching force, the Spanish commander quickly struck his colors and fled along with his 51-man garrison of troops. MacGregor announced that this was now the Republic of Florida and raised a somewhat curious flag over the fort. It was a white flag with a dark green cross. MacGregor's American recruits continued to arrive in small groups and by July, he had about 200 men to defend the island. Once the U.S. authorities realized what was going on, they prevented the rest of MacGregor's recruits from leaving the country.

By early September 1817, a Spanish force of 300 soldiers and two gunboats arrived from St. Augustine and were preparing for a counterattack. To make matters worse, Aury was late. The expected support from Aury's men wasn't there. Additionally, reinforcements from the United States weren't coming. MacGregor decided that the plan had failed. On September 4, 1817, MacGregor and most of his officers fled, taking the entire payroll with them. One of MacGregor's recruits, Jared Irwin, who was a former Pennsylvania congressman, remained as the officer in command. By then, Irwin only had 94 men remaining in the fort. Anchored nearby were three of his vessels, the

Morciana, the *St. Joseph*, and the armed schooner *Jupiter*, but Aury was still nowhere in sight.

Gunfire from the Spanish gunboats abruptly shattered the silence of a peaceful summer afternoon at precisely 3:30 on September 13, 1817. The Battle for Amelia Island was on. In the days preceding, the Spanish had erected a battery of four guns on McClure's Hill, which overlooked the fort. A few moments after the gunboats opened fire, the guns from the hill began shelling the fort. In response, the revolutionary defenders inside Fort San Carlos and those aboard the *St. Joseph* returned fire with their artillery. The exchange of fire continued until after dark. During the fight, the fort's commander, Jared Irwin, was mortally wounded and died seven days later. Even though the Spanish had only lost two men, their commander determined that he couldn't take the fort and withdrew his forces that night.

Four days later, excitement stirred when the sails of several vessels were spotted. It was Aury; he finally had arrived. There were three privateer vessels in Aury's command. His flagship was the *Mexican Congress,* which mounted twelve 18-pounder guns and had a crew of 300 men. He also had two other privateer vessels and a couple of prize vessels he had recently captured. Immediately upon landing, Aury marched into Fort San Carlos and demanded to take over as the supreme authority. He was backed up by several of his well-armed men. Irwin, who was still alive but on the verge of death, had no choice and capitulated to Aury's authority. Shortly afterwards, Aury lowered the white and green flag of the Republic of Florida and raised the flag of the Mexican Republic. Apparently Aury was still supportive of the Mexican revolution even though he was technically working for Simón Bolívar's revolution.

The opportunity of establishing a new pirate base did not escape Aury. As soon as he took control of the island, perhaps an hour after he landed, Aury began preparing the island to handle a privateer operation. His vessels began taking Spanish prizes all along the straits of Florida. Most of his prizes were slave ships. Since the United States government had forbidden the importation of slaves, smuggling them in became exceptionally profitable. The slaves Aury captured were smuggled into Georgia through Amelia Island where American slave traders paid top dollar. It is estimated that by mid-December, he had accumulated between $500,000 and $1.8 million in cash in the value of the day.

Authorities in the United States became aware of his operations shortly after he began. Even though they had absolutely no authority in Spanish territory, they came to the conclusion that Aury had to be stopped. His slave smuggling operation was in direct violation of the Act Prohibiting

Jared Irwin served as a colonel in the 5th Rifle Regiment during the War of 1812 and was elected to the House of Representatives from Pennsylvania and served from March 4, 1813–March 3, 1817.

SEPTEMBER 17–DECEMBER 23, 1817

Louis-Michel Aury operated out of Amelia Island, Florida

The Act Prohibiting Importation of Slaves took effect on January 1, 1808, and was a United States federal law that stated that no new slaves were permitted to be imported into the United States.

A corvette is a class of war ship below the class of a frigate. Determination of class depends upon size and armament. Frigates have at least 28 guns. The USS John Adams was originally rated as a frigate when it was built in 1799 but was reclassified as a corvette in 1809 to conduct patrols along the coast.

Importation of Slaves. The U.S. Navy sent a squadron to arrest Aury and stop his operations. The overall commander was Commodore J. D. Henley, and his flagship was the USS *John Adams*, a three masted corvette of war.

When the naval squadron arrived and the USS *John Adams* dropped anchor off Amelia Island's shore, the game was over for Aury. The date was December 23, 1817, and there was no battle. There weren't even any shots fired. It wasn't even a surprise. A month earlier, Aury had received communications from the U.S. government that demanded that he stop all his actions. They also informed him that the naval force was on its way to enforce those demands. After brief negotiations, Aury surrendered. U.S. naval personnel held Aury and his men for two months trying to uncover evidence that he had taken an American vessel or had smuggled slaves or goods into Georgia, but Aury was very clever. No evidence was ever found and after holding Aury for two months, he was released. Aury chose Old Providence as his next and final base.

Old Providence Island had long been associated with privateers. Located in the heart of the Caribbean, 120 miles east of the modern-day coast of Nicaragua and north of the coast of Panama, it was initially settled by English Puritans in 1629 and was used as a base for buccaneers. It was taken by the Spanish in 1641 but they abandoned it in 1670. Afterwards, English privateers including Henry Morgan used it as a base. English settlers returned during that time and remained on the island. Old Providence was claimed by Colombia during their war for independence (1808–1821), but they never really occupied it or controlled the island's administration. Even though it belonged to Colombia, the local inhabitants were mostly English and Creole.

Figure 31: *Old Providence 1774 Map*

Aury had fallen out of favor with just about everyone. Simón Bolívar's confidence in Aury was shattered. In addition to his lack of success at Galveston and Amelia Island, Aury was arrogant, pretentious, and difficult to work with. But that wasn't Aury's only problem. One by one, the revolutionary governments throughout the Caribbean began distancing themselves from privateers. They had become troublesome, and their conduct reflected poorly upon the legal status of the newly formed governments. By the end of 1820, there were very few legitimate letters of marque to be had from any of the usual sources. The only holdout was the somewhat shaky and unrecognized revolution in Buenos Aires and Chile. One of the leaders of that insurgency was Jose Cortes de Madariaga, who was issuing letters of marque to just about anyone. He even placed agents in Baltimore and Philadelphia who could issue letters directly. So Aury used the Buenos Aires insurgency as the source of his letters. Since that temporary government wasn't recognized by anyone, Aury was now technically a pirate.

Upon arrival at Old Providence, Aury immediately declared himself governor. As there was no actual standing government on the island, there was no one to oppose him. As mentioned in the previous chapter, both Jean and Pierre Laffite visited the island after leaving Galveston late in 1820. They were looking to obtain letters of marque from Buenos Aires and thought

Figure 32: *Original Buenos Ayres Letters of Marque for Luis Aury*

that Aury would be able to help them. They were probably aware that those letters were illegitimate, but they most likely believed that it gave them the pretense they needed to continue as privateers. The Laffites may or may not have personally met with Aury, although it is very likely as they finally did receive the letters of marque they requested.

Either way, Aury's presence was becoming increasingly annoying to Simón Bolívar, who wanted to get out of the privateer business and become more respectable within the international community. On January 18, 1821, Bolívar ordered Aury off of Old Providence Island. But Aury refused to leave, claiming that he was actually the governor of the island. Fate intervened and before Bolívar could take any further action, Aury was dead. He had accidentally fallen off his horse on August 20, 1821, and died at the age of 33.

But what about Aury's treasure? As you may recall, he had between $500,000 and $1.8 million in cash when he was on Amelia Island. The U.S. Navy searched the island for two months but didn't find anything. Aury certainly wasn't allowed to take it with him. And Aury couldn't secretly return to Amelia Island to recover it. The island was too busy with the inhabitants returning and the Spanish building up their fortifications after Aury left. Then, the Americans occupied the island after ownership of Florida was ceded to the United States. Perhaps the stories of buried pirate treasure on Amelia Island are true after all.

Chapter 23
Weight of the Navy

The merchant ship *Emma Sophia* was about to complete a successful voyage from Hamburg to Havana and was working her way through the Florida Straits southward along the east coast of Florida. She had just passed the Grand Bahama Banks and was approaching the Bimini Islands. That would put her position at approximately 50 miles east of Fort Lauderdale, Florida. The date was December 19, 1818, and Florida was still a Spanish colony. Suddenly, the sails of a fast-moving schooner were sighted. It was a pirate schooner with one gun and a crew of about 30 men. Shortly afterwards, the *Emma Sophia* had been captured. However, since the east coast of Florida was exceptionally well-traveled with vessels of all types sailing to and from the Caribbean, the pirates chose to take the captured ship to a nearby island far away from the busy shipping lanes. There, it could be safely looted. That island was simply identified as being somewhere among the islands of Martyr's Reef. Today, Martyr's Reef is called the Florida Keys and the *Emma Sophia* was most likely taken to Key Largo.

The Florida Keys were originally named "Los Martires" by Ponce de León in 1513. English map makers and sailors called the islands Martyr's Reef in the late 18th and early 19th centuries.

At least some of the pirates were of English descent, as the pirate who did most of the talking spoke English and went by the name of Davis. Each member of the crew was threatened with death if they didn't reveal the location of all their hidden loot. Ship's captains often hid their valuables somewhere onboard their vessel when faced with an attack by pirates. Their valuables would include gold and silver coins as well as jewelry and anything else that was worth a lot of money and could be easily hidden. In this case, the *Emma Sophia* was believed to be carrying a large number of diamonds.

A rope with a noose tied at one end was flung over the mizzen yard and William Savage, the man in charge of the ship's cargo, was brought forward. He was told that he would be hung if he didn't reveal the location of the diamonds. Savage insisted that there were no diamonds and that he had nothing of value except a pocket watch. Facing certain death, Savage saw

Robert Jacob

The *Boston Daily Advertiser* was established in 1813 as the first daily newspaper in Boston and continued to be printed until 1929.

an impractical opportunity to escape, so he leapt over the rail into the water and began to swim to shore. The pirates recaptured him and stripped off all his clothing. Then, they forced him to lay naked on the deck for about five hours. Within that time, the entire ship's cargo worth $5,000 in the currency of the day was offloaded from the *Emma Sophia* by the pirates. Afterwards, the pirates sailed away, leaving everyone alive including Savage. I'm certain that the victims were somewhat shocked and totally horrified.

That account was printed in the *Boston Daily Advertiser* on February 3, 1819, and also appeared in the U.S. Navy Proceedings Magazine, Volume 42. More importantly, it was included in a report on piracy which was written by Rear Admiral Casper F. Goodrich and presented to Congress. The United States was in the process of negotiating with Spain to take possession of Florida and the constant acts of piracy were of great concern. As Spain lost control of its colonies in the Caribbean and elsewhere in the Americas, the pirates grew bolder and far more dangerous. In addition to piracy, the smuggling of slaves into the United States was another serious issue. Recent laws forbid the importation of slaves into the United States and smuggling operations increased dramatically. In many cases, the two were connected, as pirates often captured slave ships and then took them to Texas or Florida to sell them to smugglers.

The brig USS *Enterprise*, and the schooners USS *Nonsuch* and USS *Lynx* had already been patrolling the Caribbean, but they acted independently and weren't part of an organized squadron. Closer to home, as mentioned in the previous chapter, the USS *John Adams* drove Louis-Michel Aury out of Amelia Island during the first two months of 1818. Now, after Admiral Goodrich's report in 1819, the Monroe administration established the first anti-piracy fleet with the specific mission to fight piracy along the coastline of the United States from Boston to New Orleans. The revenue cruisers *Louisiana*, *Alabama*, and *Peacock* were added to the squadron, along with several coastal gunboats. As you may recall from an earlier chapter, it was the *Alabama* and the *Louisiana* that captured the schooner captained by Jean Defarges, one of Jean Laffite's pirates, off the west coast of Florida on August 31, 1819. By 1821, six other vessels were added to the U.S. Naval Anti-Piracy Squadron. They were the USS *Hornet* with 18 guns, the 12-gun brigs USS *Enterprise* and USS *Spark*, and three 12-gun schooners, the USS *Shark*, the USS *Porpoise*, and the USS *Grampus*. These vessels primarily patrolled the waters around Cuba, but they also sailed to the eastern end of the Caribbean.

Most of the pirates in the Caribbean in 1821 were based in Cuba but there were a few other centers of piracy. The French-colonized islands of St. Martin and St. Barthélemy had long been used as bases for French privateers.

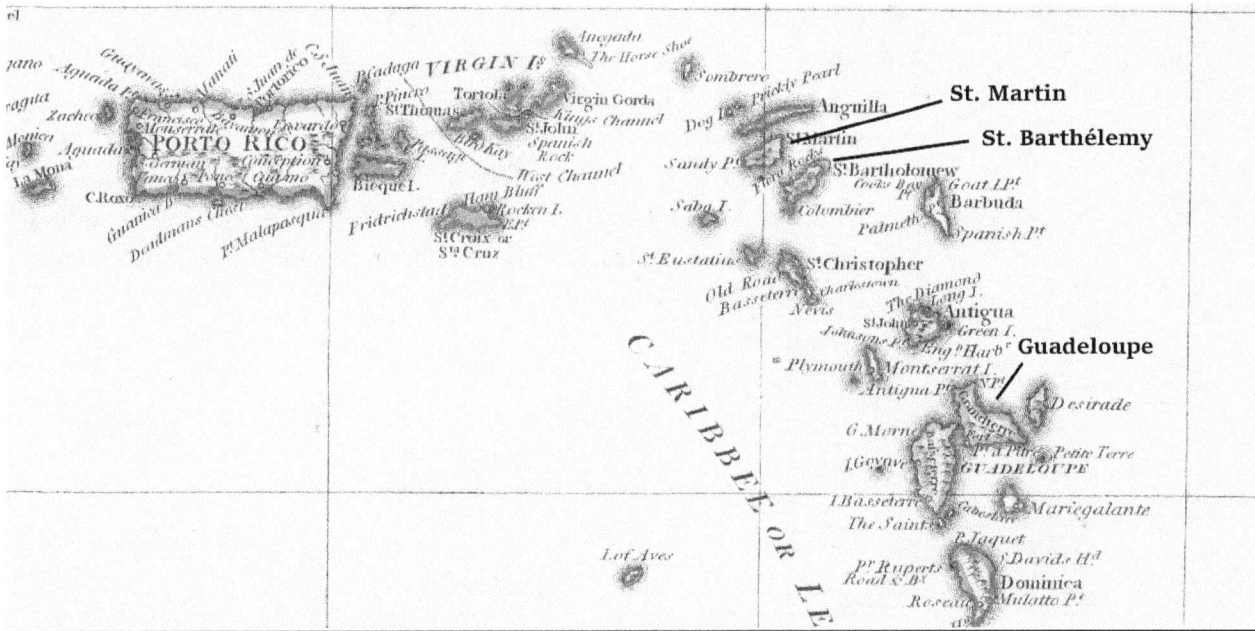

Figure 33: *Guadalupe, St. Martin, & St. Barthélemy 1798 Map*

A decade earlier, Guadalupe was the center of all French privateer activity in the Caribbean, primarily against the British, as France and Britain were at war. When the British invaded Guadalupe in 1810, the French privateers fled. Many went to Barataria to join Jean Laffite, but others established new bases on the nearby islands of St. Martin and St. Barthélemy. As the war came to a close and letters of marque became harder to get, those privateers turned to piracy. Over the decade, they were joined by a few Spanish pirates who were either anti-Spanish revolutionaries or pirates who just didn't care about global politics.

Piracy in the Caribbean was totally out of control by 1821. Thus far, the efforts of the U.S. Navy were insufficient. Most of the pirate activity was centered around Cuba because the local Spanish administration did nothing to prevent Spanish pirates from using the island for bases. As reported by the USS *Porpoise*, when U.S. Naval vessels captured Spanish pirates and turned them over to the Spanish authorities on Cuba, the pirates would be quickly released after a brief show trial. The account that sheds the best light on the Spanish administration's motives comes from the surviving victims of the brig *Cobbosseecontee*, taken by Spanish pirates within sight of Havana harbor on November 6, 1821. The pirates told their captives that the Spanish officials on Cuba supported Spanish pirates in general. They encouraged the taking of American vessels in retaliation to the United States government's support of French pirates, such as Jean Laffite, who had taken hundreds of Spanish vessels over the past fifteen years. Cuban authorities also objected to the United States government's interference with the Spanish slave trade, which was a large part of the Cuban economy.

By the end of 1821, the only colonies in the Caribbean that remained under Spanish control were Cuba and Puerto Rico. They remained under Spanish control until the end of the Spanish American War in 1898.

Isolated ports all over Cuba quickly developed as safe and secure havens for thousands of pirates. Among the most prominent were Cape Antonio, Matanzas, and Cayo Romano. The pirates who sailed the Cuban waters were all true pirates; letters of marque were no longer available from any legitimate governments in the Caribbean or elsewhere in the Americas. The time of the privateers was over, regardless of whether they were actually legitimate or just taking vessels with questionable letters of marque operating under the pretense of being legitimate.

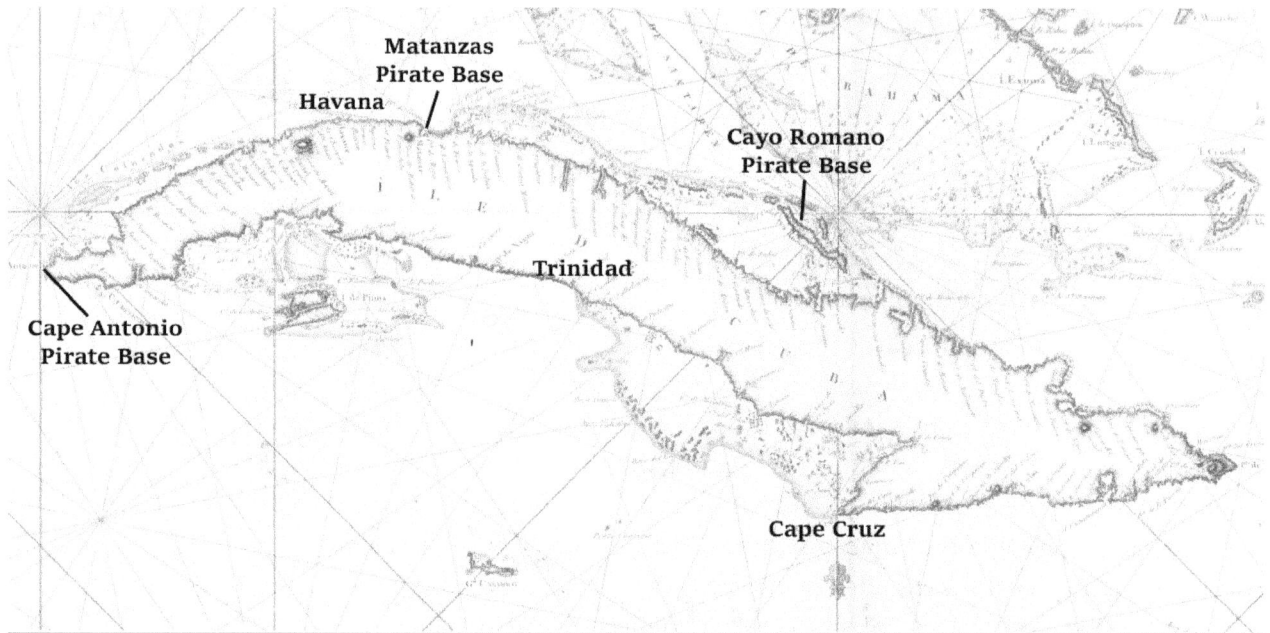

Figure 34: *Pirate Bases on Cuba 1801 Map*

Exceptionally harsh treatment of captives added to the sense of urgency for the United States in stamping out piracy. In the early 18th century, pirates such as Blackbeard and Jack Rackham seldom, if ever, actually killed anyone. Generally, the crews of captured prize vessels were well-treated. Of course, there were numerous accounts of victims being threatened and intimidated, but very few of those threats actually resulted in injury or death. Harming captives was bad for business. Pirates wanted to take vessels without a fight. If the prize crew believed that they would live through the experience and simply lose their cargo and possessions, they were far more likely to surrender without any resistance. Of course, there were a few instances of pirates torturing or killing some of their captives, but that was comparatively rare. The vast majority of 18th century pirate attacks resulted in no harsh treatment or casualties.

It was similar with the privateers of the early 19th century, only there was an added incentive for them to treat their captives well. Sailing under letters of marque technically meant that you were working for that government.

Harsh actions from privateers reflected poorly upon those governments. Privateers who gained a reputation of harsh treatment of captives could simply be denied letters of marque in the future. Additionally, with letters of marque in hand, privateers often legally sold their captured cargo to merchants in any friendly ports they visited. Bloodthirsty privateers weren't looked upon very favorably among the people who purchased their cargo. So, a reputation of kind treatment of captives was good for business. Just as with the pirates of the early 18th century, there are a few reports of atrocities committed against the captives of legitimate privateers in the early 19th century, but they are in the vast minority when compared to the reports of no harsh treatment to captives. I have not been able to find any reports of brutality committed by Jean Laffite or any of his associates when they sailed as privateers out of Barataria or Galveston.

Brig Cobbosseecontee was taken by Spanish pirates on November 6, 1821

That rapidly changed in the 1820s and harsh treatment of captives quickly became the norm. In fact, it is rare to find an account dated after 1821 that doesn't contain a significant amount of brutality. Even worse, many captured crews were all killed and their vessels burned. The men who committed those atrocities were true pirates who weren't held to any high standard of conduct. They had no letters of marque and therefore, no government to appease. They weren't going to sell their cargo to legitimate merchants. Most of the time, those pirates wouldn't even bother taking the cargo, they just took the money, jewelry, clothes, and other things they could use, then destroyed the cargo and even burned the captured vessel. The victims of the brig *Cobbosseecontee* reported that the pirates took all their clothing, cooking utensils, their sails and rigging, and several boxes of cigars they had found.

Schooner *Mary* was taken by French pirates in June of 1822

Among the most shocking of those accounts was one which came from an unnamed passenger aboard the schooner *Mary*, sailing from Philadelphia to New Orleans in June of 1822. French pirates overtook the *Mary* at dawn. In that passenger's horrifying story, he was beaten, slashed with knives, stabbed, and tied to the mast. He was forced to witness the cruelty committed upon the crew by the pirates. One crewman had both his arms cut off when he refused to reveal the location of the ship's money. The captain was tied to the deck, covered with combustible materials, and burned to death right before him. Another crewman's feet were nailed to the deck and his "body spiked through to the tiller." Other crewmen were hung on gibbets. The last member of the crew was forced to kneel with his head at the muzzle of the swivel gun. The gun was fired, severely injuring him, and causing death a few moments later. Suddenly, their brutality abruptly ended as they sighted another vessel quickly approaching. The pirates concluded their ransacking of the schooner then scuttled it so it would slowly sink. Then, they boarded their vessel and sailed away. The passenger was certain that he was destined to drown as the vessel slipped beneath the surface, but fortunately, after about two hours, the other vessel came alongside and rescued the passenger. This dramatic and vivid account was published in the *American Monthly Magazine* in the February 1824 edition, so it may be accurate, or it may be highly exaggerated by the editor using literary license to bolster circulation. Either way, it gives us an idea as to the attitude of the public and of the publishers of the time period.

There are many other accounts that appear in official reports that are not embellished through literary license. Mentioned earlier in this chapter, the brig *Cobbosseecontee* was taken by a pirate sloop within sight of Havana harbor. As the pirates ransacked the vessel, they threatened and beat members of the crew while questioning them about any hidden money onboard. In the process, the pirates hung the first mate from the maintop and drove a cutlass through the thigh of the captain.

Walking the plank was employed by Spanish pirates for some of the crew of the merchantman *Blessing* taken in May of 1822.

Walking the plank is generally believed to be simply folklore, but there was actually a documented case when Spanish pirates did make several crewmembers walk the plank. A British merchantman, the *Blessing*, was taken in May of 1822 by Spanish pirates who forced the captain and several other members of the crew to walk the plank. When they attempted to swim away, they were shot.

Captains began to fight back in retaliation. A very detailed account of a merchant crew fighting pirates came from the logbook of Captain Lamson of the brig *Belvidere* which was attacked by pirates while sailing from Port-au-Prince to New Orleans. Upon arrival at the U.S. customs inspection station on Balize at the mouth of the Mississippi, the captain made his official report and turned his logbook over to authorities. Lamson had been previously captured by pirates and was brutally treated. He wasn't about to go through that again. So, on May 2, 1822, when a suspicious looking schooner in the distance turned and headed directly toward him, he decided to make preparations to fight.

Brig *Belvidere* fought off Spanish pirates on May 3, 1822

The *Belvidere* was only armed with a 24-pounder carronade, a brass 3-pounder swivel gun, 4 muskets, and 7 pistols. All weapons were loaded and a cloth tarp was thrown over the carronade to conceal its presence. Afterwards, they waited as the schooner closed the distance. It was about 11 a.m. on May 3, 1822, when the schooner finally came within 100 yards distance and fired a single warning shot. With no reply from the *Belvidere,* the pirate schooner raised a red flag and fired a second shot. By noon, the two vessels were within hailing distance and the pirate captain shouted, "Send your boat onboard or I will murder all hands of you." The crew of the *Belvidere* could see about 20 well-armed pirates standing on deck.

Captain Lamson replied, "I will send you directly to hell." With that reply, the pirate schooner let loose a full volley of small arms fire with muskets and blunderbusses. The battle was on.

A carronade is a smoothbore cast iron naval gun with a very short barrel, generally of a large caliber, that was used in close range battle against enemy ships or personnel.

The cloth concealing the 24-pounder carronade was flung off and a few seconds later, all guns on the *Belvidere* opened fire. Six pirates immediately fell dead including their captain. A short exchange of small arms fire commenced between the two vessels. The pirate schooner had one long gun which fired three times, inflicting little actual damage to the *Belvidere*. The 24-pounder carronade fired a well-aimed shot that shattered the carriage of the pirate gun and sent it crashing into the scuppers. Some of the pirates on the schooner's deck moved to the aft quarter which offered them a better firing position. But before they could continue their fire, the *Belvidere's* brass 3-pounder, loaded with about 40 musket balls, fired directly into them and apparently killed them all in a single blast. With only about six or seven pirates left alive, the pirate schooner pulled away and was soon out of sight. The pirates had been soundly defeated. *Belvidere's* losses were only one killed and some repairable damage above the waterline.

Some success against piracy had been achieved by the navy in 1821 but it wasn't enough. While cruising off Cape Antonio on October 16, 1821, the USS *Enterprise* came upon four pirate vessels robbing three American merchant vessels. The *Enterprise* quickly intervened and captured 40 of the pirates, but the rest simply escaped to the shore. Since U.S. naval personnel were forbidden from setting foot on Cuban soil, they were unable to pursue them. Two months later, on December 21, 1821, the *Enterprise* engaged another pirate schooner, but this time the entire crew managed to escape ashore.

United States Congress authorized $500,000 to be used to fund anti-piracy operations

Congress finally authorized the funding for a larger fleet to fight pirates in the Caribbean. In 1822, a U.S. District Court was established at Pensacola to try pirates captured by naval vessels and Commodore James Biddle was placed in command of the newly designated West Indies Squadron. That squadron consisted of two frigates, the USS *Macedonian* and the USS *Congress*, two corvettes, the USS *Cayne*, and the USS *John Adams*, two brigs, the USS *Spark* and the USS *Enterprise*, four schooners, the USS *Alligator*, the USS *Grampus*, the USS *Shark*, and the USS *Porpoise*, two sloops, the USS *Hornet* and the USS *Peacock*, and two gunboats, number 158 and number 168. While some of the vessels were being fitted out, Lieutenant Matthew Perry, captain of the USS *Shark*, was sent to Key West, then called Thompson Island, to establish a base of operations for the West Indies Squadron.

The first attempt at stopping piracy in the Caribbean ended quickly in total failure. Commodore Biddle's flagship was the frigate USS *Macedonian*. It

FITTED OUT

A naval term which means to load supplies, repair damage, and do anything else required to prepare a vessel for sea.

was fitted out in Boston and sailed to join the rest of the fleet in May of 1822. Shortly after sailing, yellow fever broke out among the crew and 74 died. The USS *Macedonian* turned about and sailed to the naval base at Norfolk, Virginia for medical treatment. Biddle blamed the Boston Navy Yard's lack of sanitation for the yellow fever outbreak aboard his frigate and filed legal action. He eventually lost his lawsuit, but his required presence at the trial forced the navy to replace him with a new commander.

Commodore David Porter replaced Biddle as the new squadron's commander, which was very fortunate for the navy. Commodore David Porter was the right man for the job. Highly innovative with a true understanding of the situation he was facing, Porter totally reorganized his squadron. He knew that large vessels were unable to sail the shallower waters along the coasts of the Caribbean islands. Pirate schooners could easily escape a pursuit by sailing to those waters. Porter chose to keep only seven of the original vessels, one corvette, the USS *John Adams*, two brigs, the USS *Spark,* and the USS *Enterprise*, two schooners, the USS *Grampus* and the USS *Shark*, and the two sloops, the USS *Hornet* and the USS *Peacock*.

Porter augmented his squadron by purchasing eight smaller schooners with shallow drafts that were designed to sail the shallow waters of the Chesapeake Bay. Those schooners were similar to the type of schooners often used by the pirates. Each of Porter's new schooners was armed with three guns and assigned a crew of 31. They were the USS *Fox*, the USS *Greyhound*, the USS *Jackal*, the USS *Beagle*, the USS *Terrier*, the USS *Weasel*, the USS *Wild Cat*, and the USS *Ferret*. He also added four barges, the USS *Gnat*, the USS *Midge*, the USS *Sandfly*, and the USS *Gallinipper*. To bring supplies to his base, Porter added the transport ship USS *Decoy*, armed with 6 guns. Finally, his most innovative act was to add the steamship USS *Sea Gull*. That was the second steam powered vessel which was commissioned by the U.S. Navy. The first was the USS *Demologos*, which was renamed the USS *Fulton* by the time it was commissioned in June of 1815. The *Fulton* was actually a steam powered barge and was restricted to operation only within New York Harbor. In contrast, USS *Sea Gull* was capable of steaming throughout the Caribbean and the Gulf. Armed with three guns, the USS *Sea Gull* was fast and capable of operating in shallow water without concern for the direction of the wind.

The USS *Sea Gull* was built by the Connecticut Steamship Company in 1818. Powered by steam, it had two paddle wheels, one on each side of the vessel.

Porter named his command the "Mosquito Squadron" and assembled most of his vessels at the Norfolk Naval yard for fitting out. The Mosquito Squadron sailed on February 14, 1823, and cruised the waters near Cuba, Puerto Rico, and San Domingo before arriving at their new base on Key West on April 8, 1823. Afterwards, his Mosquito Squadron dispersed throughout the Caribbean and Gulf of Mexico. Most vessels went to Cuba, where most

Figure 35: *USS Enterprise*

of the pirates were operating. The USS *Grampus* was ordered to patrol the waters of the Gulf of Mexico and the USS *Alligator* patrolled the keys. Unfortunately for the *Alligator*, it ran aground on an unnamed reef which now bears its name. This reef is located just south of Islamorada.

The USS *Enterprise* had long been part of the anti-piracy operations. As part of the New Orleans Squadron, it captured three pirate vessels, intercepted slave ships, and supervised Jean Laffite's evacuation of Galveston in 1820. It was essential in limiting pirate activity around Cuba between 1821 and 1822. As part of the new plan of operations, the USS *Enterprise* was sent to patrol the coast of Venezuela where it struck some submerged rocks along the coast of Little Curaçao. The USS Enterprise broke up and sank on July 9, 1823, with no loss of life.

The pirate wreckers of Port Monroe on Vaca Key, Florida were among the first pirates the Mosquito Squadron eliminated. Vaca Key is located about 60 miles east of Key West and is situated between the islands of Marathon and Duck Key. The American pirates of Port Monroe were led by Joshua Appelby and were actually "wreckers." Perhaps one of the oldest methods

1823

Florida pirate wreckers were operating from Port Monroe on Vaca Key

The Mint Julep is a drink which is common in the southeastern United States and is made with bourbon, mint, water, and sugar.

of piracy, wreckers lit false signal lights on the beach, tricking any passing vessel into thinking it was actually the entrance to a harbor. Believing that it was safe to proceed, the unsuspecting vessel would run aground on a sandbar or reef. Afterwards, the wreckers would row out to the stranded vessel and loot it. That practice wasn't very glamorous or adventurous, but it was effective. Even though they were just using row boats, technically they were using vessels to capture other vessels at sea. That made them pirates. Joshua Appelby had been operating his wrecker operation for quite some time, mostly taking Spanish vessels. Now that Florida was a U.S. territory, that practice had to be stopped so Appelby was arrested.

The Mosquito Squadron initially had great success in suppressing piracy in the Caribbean during the second half of 1823. Additionally, Great Britain was faced with the same threat from pirates that the United States was. After Commodore Porter took command, the two governments decided to cooperate with each other and the naval forces from both nations began coordinating their efforts. However, the success of 1823 was followed by marginal success in the following two years, primarily due to a yellow fever epidemic throughout the fleet. Porter was forced to pull most of his vessels back to receive medical treatment and in their absence, piracy around Puerto Rico made a dramatic comeback.

Commodore Porter himself fell ill with the dreaded disease while he was at Key West. Medical physicians of the day usually placed all yellow fever patients on a very restrictive diet, allowing them to consume only the food and drink that they deemed safe. That treatment is still commonly practiced by physicians today. In Commodore Porter's case, his doctors were certain that he was on the verge of death and granted him permission to eat or drink whatever he wished. It was sort of like granting him his "last request." Porter immediately ordered mint juleps which were made with extra bourbon. The bedridden naval officer drank one mint julep after another. To everyone's astonishment, his condition gradually began to improve and within several days he had made a full recovery. Commodore Porter always attributed his recovery to those mint juleps.

In October of 1824, the USS *Peacock* under the command of Lieutenant Platt returned to Puerto Rico. American merchants informed him that $5,000 worth of American goods in the value of the day had recently been taken by Puerto Rican pirates were being stored near the town of Fajardo on the west end of the island. The Lieutenant decided to retrieve the merchandise and led a small force that landed ashore. However, the Puerto Rican authorities intervened and arrested him. Commodore Porter couldn't tolerate such an offense against one of his officers and landed a large military force on Puerto Rican soil which rescued Lieutenant Platt. Porter's actions caused a major diplomatic incident and Porter was court-marshaled and relieved of command.

That incident actually enlightened the United States Congress as well as President Monroe to the reality of the pirate crisis. Cuba and Puerto Rico remained the only two Spanish colonies in the Caribbean. Monroe suggested that the United States should blockade Cuban and Puerto Rican ports if they continued to support pirates. He also suggested that U.S. naval forces should be allowed to pursue pirates onto Cuban and Puerto Rican soil. Meanwhile, Spain finally accepted the inevitable and officially recognized the revolutionary governments in Central and South America, relinquishing all control. Those newly recognized governments immediately ceased all attacks against Spanish shipping. With everyone at peace, the Spanish pirates began taking Cuban and Puerto Rican prizes as well. The combination of the possibility of conflict with the United States and the escalating attacks on their own merchant vessels forced the Spanish government to change its position. Spanish authorities began arresting and executing pirates wherever they found them.

Success for the U.S. Navy was on the rise again and this time the Spanish authorities were cooperating with them. Working in coordination with two Spanish sloops, the USS *Grampus* engaged the pirate schooner *El Mosquito* on March 2, 1825. That was the schooner of the famous Puerto Rican pirate Roberto Cofresi. The *El Mosquito* was disabled and the captain and crew escaped ashore. But this time, they were arrested by local authorities. Cofresi and his men were executed by firing squad on March 29, 1825.

Coordination between British and American naval forces began in earnest during March 1825. Previously, the two navies had cooperated in pursuing pirates, but now, they actually planned and coordinated an operation as a

Commodore David Porter left the U.S. Navy in 1826 and became the Commander-In-Chief of the Mexican Navy. The Andrew Jackson Administration recalled him and appointed him minister to the Barbary States and later as Chargé d'Affaires to the Ottoman Empire.

By 1825, Spain had conceded to the revolutionaries and relinquished control of all their colonies in South and Central America except Cuba and Puerto Rico.

Figure 36: *USS Grampus*

combined force. The USS *Sea Gull* and the USS *Gallinipper* joined forces with the HMS *Dartmouth*, HMS *Lion*, and HMS *Union*. On March 25, 1825, they attacked pirate vessels along the north coast of Cuba.

The pirates in the Caribbean and the Gulf of Mexico had felt the full weight of both the U.S. Navy and the British Royal Navy. During the anti-piracy operations of the 1820s, U.S. naval forces engaged approximately 3,000 pirates, capturing 79 vessels, and arresting 1,300 pirates. The Royal Navy engaged 13 pirate vessels and arrested 291 pirates. And in the final year, the Spanish Navy managed to catch 5 vessels and 150 men. By the end of 1825, piracy in the Caribbean and the Gulf of Mexico was all but eradicated.

Chapter 24
Florida's Pirate Legend, José Gaspar

For Floridians, José Gaspar is the most famous pirate in the entire history of Florida and perhaps even the United States or conceivably the entire world. Every January for well over 100 years, Tampa becomes inundated with a frenzy of pirates. It's called the Gasparilla Pirate Invasion and the entire downtown area is magically transformed into a pirate-themed expanse. This huge and well-planned event has a boat regatta with pirate ships, children's events, dances, formal balls, and of course several enormous parades comprised of many pirate floats and lots of plastic beaded necklaces. These parades easily rival any of the Mardi Gras parades held in New Orleans.

José Gaspar himself makes an appearance and reenacts the traditional capture of the city. Well, actually it's a person pretending to be José Gaspar. The host group of the Gasparilla Pirate Invasion has always been Ye Mystic Krewe of Gasparilla, who has adopted José Gaspar the pirate as their "patron rogue of their city-wide celebration." The third largest parade in the nation, it is estimated that each year the combined Gasparilla events bring in about $40 million to Tampa's economy. That's a lot more money than José Gaspar ever supposedly took.

Who was José Gaspar and was he a real pirate? Those questions are unimportant to the vast majority of the participants of the Gasparilla festivities. They attend the events to have a great time, dress in pirate costumes, and to allow themselves to be swept up in the imaginary world of the Gaspar legend. Those questions are also unimportant to the many business owners who depend upon the success of Gasparilla events to make a living. But for historians and pirate enthusiasts who search for the truth behind the legend, those questions rise to the surface among the deluge of media and publicity which promotes the Gasparilla Pirate Invasion. Those questions assume monumental importance.

Most enlightened historians agree that José Gaspar did not exist in real life, at least not as he is portrayed in the commonly circulated stories and legends. There is evidence that there was an actual pirate named José Gaspar, but he never set foot on Florida soil. Sometimes, the story of how a legend emerges is more fascinating than the legend itself. That is certainly the case with the real story of how Florida's José Gaspar came to life. The origins of how the legend developed are just as complex and intriguing as the actual legend of José Gaspar. But before we can delve into the world of historical fact and uncover the truths behind the myth, we must first become familiar with Gaspar's legend as it currently stands. That's easier said than done, as there are many versions of his legend, especially of his earlier years. Some even completely contradict each other. Therefore, I have chosen to present a version of Gaspar's story that combines the original written source with additional accounts from other versions that make the most logical sense. For the moment, let us imagine that José Gaspar was real.

Imagine that José Gaspar was real

Born in Seville, Spain, in 1756, possibly to nobility, the young José found himself in a great deal of trouble as a teenager. He became infatuated with a beautiful young woman who didn't return his affections, so he chose to kidnap her. He was quickly arrested, and the judge gave him the choice of prison or service in the navy. José chose the navy and because of his family's influence, he was appointed to the Spanish Naval Academy. Upon graduation, José's naval squadron was sent to fight the Barbary pirates in the Mediterranean Sea. During the 1770s, he quickly rose through the ranks and was promoted to captain. His squadron was then sent to the Caribbean in the early 1780s where he distinguished himself further by capturing a large number of pirates. In 1782 he was rewarded by promotion to admiral and assigned to the court of King Carlos III as his naval attaché.

José had difficulty at court. He wasn't used to the political backstabbing that was common and quickly made political enemies with a court official named Manuel Godoy. Additionally, he showed poor judgment with regard to the ladies who were ever-present at court. Throughout his entire life, José always exhibited a weakness for attractive women. Shortly after his arrival, he became romantically involved with the king's daughter-in-law, Maria-Luisa. However, he also fell in love with a woman named Dona Rosalita, whom he subsequently married. The jilted Maria-Luisa was furious and vowed revenge. Eventually she found an ally in the person of Manuel Godoy.

King Carlos III was king of Spain from 1759–1788.

José and Dona Rosalita had two children by 1783, one boy and one girl. Since they had only known each other for about one year, one must assume that his children were twins. Meanwhile, Manuel Godoy and Maria-Luisa were plotting to destroy José. The crown jewels needed to be transported by naval vessel and the job was given to José Gaspar and his ship, the *Flora Blanca*. For Manuel Godoy, this was his chance for revenge. He secretly hired members of Gaspar's crew to steal the jewels and make it look as if Gaspar was to blame. But that wasn't enough. Godoy's men raided the home of José and raped and killed his wife and even killed his infant daughter. King Carlos III believed Godoy when he told him that Gaspar had stolen the crown jewels. A royal decree was issued which put a price on Gaspar's head. When the news reached Gaspar, he was furious. He took on a new crew of pirates and swore eternal vengeance upon all Spaniards in general. He left Spain on the *Flora Blanca* and sailed for America.

The *Flora Blanca* had just rounded the Florida Keys and was sailing north along the west coast of Florida when José spotted the ideal location for a pirate base. It was an island with long beaches and fresh water that shielded a small bay where ships could be hidden. The bay was connected to the Gulf by a deep channel, perfect for entering and leaving the bay safely. Additionally, there were many small islands nearby that could be used for storage, holding captives, and to house his growing crew of pirates. The bay he spotted was Turtle Bay and the island was Boca Grande. Gaspar chose Boca Grande as his new home and named it after himself, Gasparilla Island. Gasparilla means Gaspar the Outlaw, a nickname he often used.

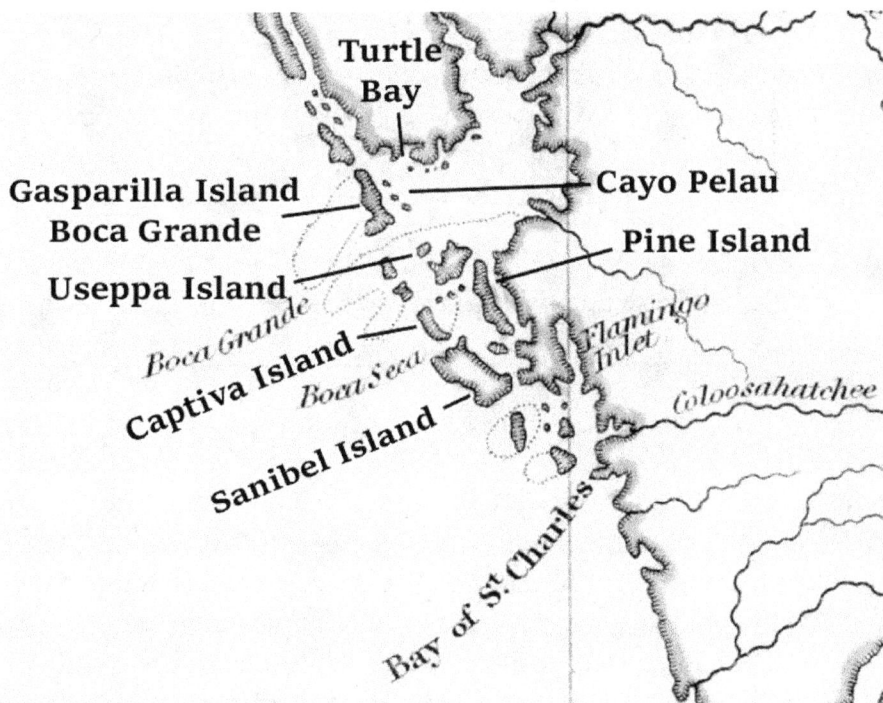

Figure 37: *Boca Grande from an 1860 Map*

An exquisite house was eventually built on Gasparilla Island where Gaspar lived. His crew built a pirate village on the island of Cayo Pelau which lies a little more than one mile to the east. Cayo Pelau had an enormous mound in the center that measured about 50 feet at its highest. It had been constructed by the Calusa as a burial mound hundreds of years earlier. Gaspar ordered a lookout tower to be built on the top of the mound. From there, passing ships could be easily spotted. Once finished, the lookout tower was manned by one of the crew at all times.

Gaspar was an exceptionally cruel pirate. He killed all the male passengers of every vessel he took. Then, he would line up all the women and children. After separating the attractive women from the rest, the unattractive or old women were thrown to the sharks along with the children. All of the remaining female captives would either be given to his crew or kept for himself. The females were taken to an island set aside for the sole purpose of housing and guarding them. They were imprisoned in a stockade of twelve houses constructed of palmetto logs and arranged in a semi-circle close to the water's edge where they were kept under constant guard. However, their stay may have been a short one. There was only room for a certain number of women captives. Each time a new prisoner was added to Gaspar's harem, he would choose one to die. This island was eventually named Captiva Island for the captives that were ever-present during Gaspar's time.

The tradition of killing all male prisoners was briefly interrupted in September 1801 when Gaspar took the Spanish vessel, *Villa Rica*. Aboard was a man who wished to join the pirate crew and for some unknown reason Gaspar agreed to take him on. His name was Juan Gómez, and he would become the prime source of much of the information we have about José Gaspar. The ship was taken about 40 miles from Gasparilla Island and had on board a cargo of chests made of copper which contained large amounts of gold. Of even more interest to Gaspar, the ship was carrying twelve beautiful young women.

Several months earlier, a Spanish princess had traveled to Mexico where she was royally entertained and befriended eleven of the fairest daughters of the Mexican nobility. Some say that she was in fact Josefa de Mayorga, the daughter of the Spanish Viceroy, Martín de Mayorga. The princess was taking the eleven young ladies back to Spain where they could be educated in Spanish customs. Now, they were at the mercy of the brutal José Gaspar. The charming and exceptionally attractive Spanish princess, whom Gómez called "the little princess," proved irresistible to Gaspar and he immediately fell in love with her. However, she refused all of his affection and spurned him. Her rejection infuriated Gaspar and he fell into a fit of rage. Suddenly, Gaspar drew his sword and cut off her head. Immediately afterwards, he felt exceptionally remorseful and decided to bury her with honors. He took her to an uninhabited island close to his base and buried her there. He named the island in her honor, the island of Josefa. In the subsequent years, the name Josefa Island morphed into the name Useppa Island.

Jean Laffite was among the most successful pirates of the early 19th century, and it isn't surprising that José Gaspar and Jean Laffite eventually met and became friends. What is surprising is their secret assignment for the United States Government. In 1803, through secret negotiations, President Thomas Jefferson arranged to purchase Louisiana from France. Known as the Louisiana Purchase, the deal involved a down payment of $3 million in gold to be secretly carried to France. Wishing to keep the details of the Louisiana Purchase as quiet as possible, the job was given to José Gaspar and Jean Laffite. Afterwards, Gaspar maintained a friendly relationship with Laffite, often visiting Barataria and selling some of his captured merchandise to New Orleans merchants.

Black Caesar was another famous pirate who befriended Gaspar. Originally sailing from his base on Elliott's Key, Henri "Black" Caesar established a residence on Sanibel Island that he used while he was sailing with Gaspar. Some say that Sanibel Island was named for the true love of Rodrigo Lopez, Gaspar's first mate. Others say Sanibel was a woman who Gaspar himself loved and lost. Either way, Black Caesar and José Gaspar enjoyed great success while sailing together. Gaspar kept his loot buried on Boca Grande while Black Caesar tended to bury his loot on several of the nearby islands and throughout the keys. Others soon joined the pirate kingdom. One was Brewster Baker who was an Englishman and had led a mutiny aboard his merchant vessel and turned pirate. He built his home on the north end of Pine Island at a spot named Bokeelia. Gaspar was also joined by another pirate known as Old King John. By 1819, Gaspar's pirate kingdom had grown to about 200 men.

When the 1819 treaty to make Florida part of the United States was signed, many of Gaspar's pirates began to question whether they should remain on the Florida coast. Realizing that they would no longer be sailing in Spanish waters and living on Spanish soil, they knew that the United States would not tolerate their actions. Gaspar's entire organization called for a meeting which was held on Sanibel Island. After a long discussion, most of Gaspar's pirates voted to leave. Among those who left were Brewster Baker, Old King John, and Henri "Black" Caesar. José Gaspar was once again sailing alone with a single vessel, the *Flora Blanca* and his loyal crew.

However, Gaspar wasn't alone for long. In May of 1820, Jean Laffite was forced to leave his pirate base on Galveston and was looking for a new base from

Late 19th century maps show the island as being named Useppa Island, but that same island is clearly named Josefa Island on an 1833 map.

Henri "Black" Caesar operated from his base on Elliott's Key, Florida from 1805–1818.

SEPTEMBER 1817–MAY 1820

Jean Laffite operated a pirate base on Galveston

which to operate. Sailing with a single vessel and a crew of about 30 men, Laffite dropped anchor off Boca Grande and rowed ashore to speak with his old friend José Gaspar. Since the rest of his pirate friends had abandoned him, Gaspar was delighted to welcome his old friend aboard.

By the close of 1821 it was becoming more and more difficult to find and take prizes. Gaspar couldn't rely on the usual tactics of simply waiting for a prize vessel to sail by. He realized that he needed inside information in order to target rich prizes. Gaspar managed to pay a government official to pass information on to him concerning the secret shipment of gold. His spy came through and informed him about a shipment of gold worth $100,000 in early 19th century currency that was being taken from Havana to New Orleans aboard the steamship, *Robert Fulton*.

Robert Fulton invented the steamboat in 1807.

Gaspar knew that he could never catch a steamship, so he devised a clever and bold plan. He sailed twelve small vessels to Cuban waters, perhaps ones he had recently captured, and ran them all aground on a sandbar near the northern coast of Cuba. His personal schooner the *Flora Blanca* remained close by but out of sight. Gaspar's plan was to give the appearance of being shipwrecked. As the *Robert Fulton* steamed by, Gaspar was certain that the steamer would come to their rescue and bring his pirates onboard. Then, his pirates could overwhelm the crew by surprise and capture the steamer. Once securely in Gaspar's control, he would signal the *Flora Blanca* to come alongside. The pirates could then transfer the entire gold shipment to the schooner and safely sail away.

At first everything seemed to be working well. The *Robert Fulton* saw the stranded vessels and changed course towards them. But as the steamer came within shouting distance, one of Gaspar's men who had issues with Gaspar's leadership revealed the deception and the *Robert Fulton* abruptly turned and steamed away. It is not known what became of the disloyal crewman who betrayed Gaspar.

By the spring of 1822, the U.S. government had begun its war on piracy and to the 80 remaining pirates on Gasparilla Island it seemed that the time was right for them to retire. José Gaspar, Jean Laffite, and all remaining pirates decided to divide all their treasure and live out their lives in luxury ashore. They had approximately $30 million in today's currency hidden in six locations around the island of Boca Grande. But before they had the opportunity to gather any treasure for division, a loud and excited shout came from the lookout. "Ship Ahoy! It be a British merchant brig." The lookout had spotted a large merchant brig flying a British flag passing close to their pirate base. Gaspar couldn't resist taking such an apparently easy prize that was sailing so close by, so he gave the orders to attack.

Gaspar's 80 pirates were divided into three groups. Thirty-five men went with Laffite on his schooner, thirty-five men went with Gaspar on the *Flora Blanca*, and the remaining ten men were ordered to stay on the island to guard the treasure. Juan Gómez was among the ten pirates left on shore. Gaspar's pirates proved to be well trained and experienced. They dashed from the beach and climbed aboard their two vessels. Scampering up the rigging, the pirates quickly raised the sails and got their vessels underway in record time. In what seemed like only five minutes, the two pirate vessels were sailing swiftly through Boca Grande Pass and straight toward the merchant vessel.

As they approached, Gaspar ordered his gunner to fire a warning shot across the brig's bow. A blast came from one of the guns on the *Flora Blanca* and as the cannon ball splashed in front of the unidentified vessel, the British flag was lowered. That was the sign of surrender and Gaspar most likely thought that this prize would be taken easily. He was wrong. A moment later, the stars and stripes of the United States was hauled up and the gun ports of a warship were flung open. The unidentified vessel was actually the USS *Enterprise*. Gaspar and his crew stood on the deck in a state of shock. A moment later, the guns of the *Enterprise* fired a well-coordinated and devastating broadside. For Gaspar, it must have seemed as if the world blew up in his face.

The *Flora Blanca* returned fire, but it couldn't match the relentless fire from the *Enterprise*. After the brief battle, the *Flora Blanca* was riddled by cannon balls and was taking on water below deck. Meanwhile, Jean Laffite watched in total astonishment from his schooner that remained safely out of the fight. Gaspar knew that his vessel was lost. He also knew what awaited him if he was captured. In a last moment of defiance, José Gaspar tied an anchor chain around his waist and shouted, "Gasparilla dies by his own hand, not the enemy's!" After those famous last words, Gaspar leaped into the dark blue sea and sank into oblivion.

The members of the pirate crew of the *Flora Blanca* who survived the fight were all captured and hung from the yardarm except for the sixteen-year-old cabin boy, who happened to be the son of Juan Gómez. Seeing that the battle was lost, Jean Laffite came about and sailed northward. The next day, however, Laffite was spotted by another pirate hunter near the mouth of the Manatee River. After a short battle, Laffite's schooner began to sink. Laffite and a few of his crew managed to make it to shore where they faded into the dense jungle, never to be heard from again.

Juan Gómez is the prime source of the details of Gaspar's piratical deeds in Florida. As one of the ten pirates who remained on Boca Grande during

Mauritius Island is located in the Indian Ocean and was under French administration from 1715 to 1810.

Gaspar's battle with the *Enterprise,* he escaped capture. Apparently, those ten pirates didn't know the precise locations of Gaspar's buried treasure, or they would have taken it with them. Juan Gómez stayed in Florida for the rest of his exceptionally long life. He lived to be 119 years old and recounted the details of his service as a member of José Gaspar's crew to tourists and newspaper reporters many times.

Of Portuguese heritage, Juan Gómez was born on Mauritius Island in the year 1781. His family relocated to France during the early reign of Napoleon Bonaparte, taking Juan with them. In September 1801, he was sailing as a crewman aboard the Spanish vessel, *Villa Rica,* when it was taken by José Gaspar. It isn't clear why Juan Gómez was allowed to live and join Gaspar's crew instead of being killed, as was Gaspar's custom, but shortly afterwards, Juan was promoted to one of Gaspar's first mates. It has been suggested that Juan was actually José Gaspar's brother-in-law, although there is no mention of Gaspar having a sister or wife. While serving with the pirates, Juan fathered a son, who became Gaspar's cabin boy. Juan Jr. was arrested after the capture of the *Flora Blanca* and imprisoned for ten years. He died in Florida in 1875, twenty-five years before the death of his father.

After the defeat of Gaspar, Juan wandered into the Florida wilderness and carved out a modest living for himself doing a variety of jobs over the next 78 years. Gómez served in the U.S. Army under Colonel Zachary Taylor during the Second Seminole War, where he anglicized his name to John. Eventually, he settled on the island of Panther Key, which is among the "Ten Thousand Islands." As a longtime resident of that island, he became known as Panther John. In the 1880s, as Florida's tourist industry began, Panther John made a living as a fishing guide. He would often tell his customers the stories of his time as a pirate with José Gaspar. Panther John also gave several interviews to members of the local press in the 1890s. Unfortunately, in 1900, the still very active Panther John Gómez became tangled in his fishing net and drowned. The last member of Gaspar's crew had finally met his end.

The Ten Thousand Islands are a chain of islands and mangrove islets off the southwest coast of Florida between Marco Island and the mouth of the Lostmans River. Despite the name, the islets only number in the hundreds. They have been a popular fishing area for tourists since the 1880s.

Birth of the Legend: The Truth Behind the Myth

There actually was a pirate named José Gaspar, but it is very doubtful that he ever visited Florida. A very interesting article appeared in the September 15, 1821, edition of the *Floridian,* which was a Pensacola newspaper, published by Nicholas & Tunstall between 1821 and 1824. The article gives the details of a pirate named José Gaspar who sailed from his home port on the pirate friendly island of St. Bartolome. Gaspar was using worthless letters of marque issued to him by the revolutionary government of Uruguay. On

August 19, 1821, while cruising the waters of the eastern Caribbean aboard his schooner *Jupiter*, Gaspar captured two American vessels from New Orleans and held them for two days. This was confirmed months later by a naval report which stated that the USS *Grampus* tracked down Gaspar and captured his vessel in October of 1821.

As far as Florida's José Gaspar is concerned, there is absolutely no evidence that he ever lived. There are no reports of a pirate named José Gaspar taking a large number of vessels along the southwest coast of Florida in the early 19th century. There aren't even any reports of a large number of vessels being taken by any pirate along the southwest coast of Florida in the early 19th century. Similarly, there are no reports from any U.S. naval vessels of any engagements or pursuits of pirates near Boca Grande anytime in the 1820s. Certainly the USS *Enterprise* never engaged a pirate vessel as described in the legend. The logbooks would have clearly described the encounter, but a thorough search of the *Enterprise's* logbook reveals that no action off the Florida coast ever occurred and that no pirate named José Gaspar was killed or his crew arrested. The USS *Enterprise* wasn't even in Florida waters between the fall of 1821 and the spring of 1822, it was cruising the waters off the southern coast of Cuba. Finally, dozens of historians have searched the Spanish records of the late 18th and early 19th century looking for anyone named José Gaspar. They came up empty. The Spanish kept meticulous records and if someone named José Gaspar had attended their naval academy or even had served as an officer in the navy, they would have recorded it.

It seems as if the real birth of José Gaspar's legend actually came from an 1898 article written to promote the 1904 Louisiana Purchase Exposition, informally known as the St. Louis World's Fair. As is evident by the name, this exposition celebrated the 100th anniversary of the Louisiana Purchase. However, in 1898 the location of this monumental event had not yet been chosen. There were several cities that were in stiff competition for the honor of being the host city. Even though the construction of the exposition buildings was still five years away, the city planners of St. Louis wanted to get a head start and make sure that their city would be selected as the host city. Of course, St. Louis already had a great connection to the Louisiana Purchase as it was very near the spot where the Lewis and Clark Expedition started its journey in 1804. But to the planners in St. Louis, that might not be enough to convince the decision makers and the financial backers to select St. Louis as the location for the exposition. So, a publicity campaign was launched promoting St. Louis as the best choice. This campaign included lots of memorabilia, posters, and flyers as well as several stories and articles on the history and importance of the Louisiana Purchase. Some of those stories and articles were historically sound, while others were works of highly imaginative fiction. Among the most interesting of those fictional

The Floridian, was a Pensacola newspaper published from 1821–1824.

The Louisiana Purchase Exposition, informally known as the St. Louis World's Fair, was an international exposition held in St. Louis, Missouri from April 30 to December 1, 1904.

stories was one that involved the pirates Jean Laffite and José Gaspar. Here's how the story went.

Thomas Jefferson had negotiated the Louisiana Purchase with France but was still waiting on the final approval from congress. As part of the deal, the United States had to make a down payment of $3 million in gold directly to the French government. Jefferson knew that this payment had to be delivered in the utmost secrecy. Congress hadn't officially committed to the financial arrangements and if news leaked out that the down payment had already been sent to France without congressional approval, the opposition party could use that information to squash the entire purchase or even destroy Jefferson's presidency. For reasons of political security, the carriers of this gold had to be far removed from any office within the U.S. government. Jefferson's agents decided to use pirates to transport the gold to France and managed to make contact with two of them who agreed to accept the job. They were of course Jean Laffite and José Gaspar.

The story was filled with a great deal of political intrigue, adventure, and government conspiracy which always has a certain amount of public appeal. Although there was absolutely no truth to that story, the general public who read it loved it. In truth, there was no down payment of gold. France agreed to allow the United States to assume some of their international debts instead of sending the down payment. Additionally, it is highly doubtful that the government would have trusted $3 million in gold and the ultimate success of the Louisiana Purchase to a couple of pirates.

Jean Laffite was a terrific choice as one of the main characters. The most famous pirate of the 19th century, the use of his name gave the story a sense of legitimacy. The author must have counted on the readers not knowing that in 1803, Laffite was totally unknown and wouldn't have been chosen by any of Jefferson's agents to take on such a mission. It isn't clear how the author came upon the name of José Gaspar as the partner in this adventure. Perhaps that name came from fictional accounts of pirates known to the author or perhaps it was still circulating within the public memory from the events of the real José Gaspar, the pirate from St. Bartolome. Either way, José Gaspar's name was suddenly thrust into the media and will be forever linked with that of Jean Laffite.

Meanwhile, John Gómez was making a living as a fishing guide on Marco Island, Florida. Since the late 1870s, the Florida tourism industry had been flourishing, especially when it came to fishing. The Ten Thousand Islands area of Florida was a favorite spot for wealthy yachtsmen who were vacationing from the north. They would often hire fishing guides and John Gómez was supposedly one of the best. But in addition to good tips on fishing, he provided bloodcurdling stories about his days as a pirate. The tourists loved it, and his business took off. Shortly afterwards, John discovered that the more gullible tourists would even buy treasure maps, so he began telling tales of buried treasure and producing fake treasure maps which he would sell to them.

Gómez continued to develop his business as a fishing guide as well as a teller of pirate stories and seller of treasure maps. He eventually realized that he had to invent a more colorful past to go along with his pirate stories. By taking a look at the historical record, one can see a change in John Gómez in the late 1870s as he began to invent a pirate background for himself. In the 1870 census, John gave his birth date as 1828 and didn't list his country of origin. A birthdate of 1828 would have made him 72 years old when he died, far more believable than the many other birth years provided which would have made him over 120 years old at his death. But by the 1880 census, John was claiming that he was born in France in the year 1785. That changed to Corsica by the 1885 census. Subsequent censuses recorded his birth year and birthplace as 1778 in Honduras, then in 1781 in Portugal, and then in 1795 in Mauritius. His last Census of 1900 listed 1776 as his birth year and gave his birthplace once again as Portugal. That would put his age at 123 years old at his death.

Panther Key is among the Cape Romano Ten Thousand Islands chain. It is only accessible by boat. The island is uninhabited today but was inhabited by John Gómez and a few others in the late 19th century.

In 1884, Gómez settled on Panther Key and took on the colorful name of Panther John Gómez. He was an exceptionally well known and loved character among the locals. His ability to tell an exciting story had made him a celebrity. While there is no doubt that he told many a tale about pirates and their buried treasure, it isn't certain that he ever mentioned the name of José Gaspar. He may or may not have identified himself as one of Gaspar's pirates—historians can't be certain either way. There are no newspaper stories that were published in the first years of the 20th century that mention Panther John as being a former member of José Gaspar's crew or that he ever claimed to have known Gaspar. So far, no source document has been found that clearly links Panther John Gómez to José Gaspar. That connection came later, after his death.

In 1901, Ricardo the Mystic was the next person to contribute to the birth of the José Gaspar legend. The leader of a group of entertainers called Ricardo the Mystic and Company, he traveled the Vaudeville circuit performing magic shows and hypnotizing members of the audience. At the time, hypnotists were all the rage. While in Tampa, Florida, he heard of a story of a pirate named Gaspar who had buried some treasure nearby. Where and how he heard of that story are unknown. Looking to make some fast money, Ricardo the Mystic began his own personal séance fraud titled "Find the Gold of Gaspar." Ricardo's unsuspecting and highly gullible victims would pay him to attend a fake séance where Ricardo would supposedly contact some of Gaspar's dead crewmen and possibly even Gaspar himself. Ricardo told his victims that once contacted, the spirits may possibly reveal the secret location of their treasure. Eventually the scam was uncovered.

Ye Mystic Krewe of Gasparilla

The random collection of tales, rumors, and speculation about the existence of José Gaspar leapt into the pages of history with the creation of Ye Mystic Krewe of Gasparilla in 1904. The city planners of Tampa were making preparations for the city's first May Day Festival. They hoped that the event would possibly develop into a festival as big as Mardi Gras in New Orleans and bring in millions of dollars to boost the local economy. However, the event would have to have something special in order to be a success. The planners needed an exciting and spectacular event to culminate the day's festivities. Since theatrical pirates all dressed up in lavish costumes are always exciting and entertaining, they decided to end the May Day Festival with a magnificent pirate dress ball.

Secret meetings were held to plan the surprise pirate events and Ye Mystic Krewe of Gasparilla was founded. The original forty members of the Krewe planned a horseback invasion staged at the Tampa Bay Hotel, a pirate parade through town, and the first Coronation Ball where the pirate King and Queen would be crowned. But most importantly, the Krewe needed a figure to base their organization on. They needed a real pirate from Florida. One of the Krewe's members was Edwin D. Lambright, a professional writer and editor of *The Morning Tribune*, a local Tampa newspaper. He was chosen to bring the legend of Gasparilla to life and continued to add more details to the story over the next several years.

The event was a huge success. A long and very descriptive story about the first festival ball appeared in *The Morning Tribune* on May 10, 1904, titled *"How They Disported Themselves in the Festival Ball—An Account of the Swellest Function Ever Given in the City."* The story suggests that Gasparilla was known to locals before the event by stating ". . . others who have been longer in the State and have heard of Gasparilla [and] perhaps know the story of the bold pirate Prince who came with his adventurous band from their home in Spain in search of excitement and wealth." The first pirate king of the ball was referred to as Edward Gasparilla, in reality, he was Edward Roach Gunby, a local political figure and lawyer. But in those early years, the full name of José Gaspar wasn't used in conjunction with the Tampa festival, at least not in print. He was always referred to as simply "Gasparilla."

The pirate events continued to grow as part of the May Day festival until 1913 when the city planners of Tampa decided to transform the festival into

The Tampa Tribune was established in 1895 and is still in operation. It has been published under the names of *The Morning Tribune, The Tampa Morning Tribune, The Tampa Sunday Tribune,* and *The Tampa Daily News.*

a full-scale standalone pirate-themed event. They also changed the month of the event from May to February. An article announcing that change was published in *The Tampa Morning Tribune* on February 1, 1913, titled *"Gasparilla Day to be Greatest Event of its Kind for the City."* In the article the author wrote that it "may rival in magnitude the Mardi Gras of New Orleans.

Charlotte Harbor & Northern Railway

The final step in the process needed to propel the legend of José Gaspar into the status of immortality was a well-funded publicity campaign which produced promotional material featuring José Gaspar. That materialized in 1913, the same year that the Gasparilla Pirate Invasion became a standalone event. That publicity campaign was brought about through the imagination and vision of Robert S. Bradley, President of the Charlotte Harbor & Northern Railway Company. But what does a railway company have to do with Gasparilla?

Sixty laborers from the Peace River Mining Company and the Army Corps of Engineers landed on Gasparilla Island in 1905. They began the construction of a phosphate mine. Peter B. Bradley, a chief executive of the American Agricultural Chemicals Company, purchased the mine that same year. He also purchased a railroad line and began construction of a railway bridge and spur that connected the island to the mainland. Completed in 1907, the railway enabled supplies to be brought in for the miners and for the phosphate to be shipped out. The railway was renamed the Charlotte Harbor & Northern Railway Company and Peter's brother, Robert S. Bradley, became the railway's president. So taken by the beauty of the island, the Bradleys formed the Boca Grande Land Company in order to capitalize on the real-estate potential as well. By 1909, a new upscale community was planned along with a luxury resort hotel that was named the Hotel Boca Grande. It opened in 1911. At first, the hotel was supposed to be for company officers only, but the Bradleys quickly realized the potential tourist market, so in 1912, the hotel was expanded into a world-class resort that would be open to anyone who could afford it.

Robert Bradley knew that the resort itself wouldn't be enough to attract customers; he needed a promotional marketing stratagem to make his resort unique and special. His timing was absolutely perfect. That new resort was scheduled to open for the 1913 season, at precisely the same time that Tampa's new Gasparilla Pirate Invasion was scheduled to début. Bradley realized that this pirate invasion was destined to become the city's premier attraction and he wanted his new resort on Gasparilla Island to capitalize on it. The success of both would mean big profits for the railway. Thousands of tourists would buy railway tickets to attend the pirate invasion in Tampa. Others

First chartered in 1897 as the Alafia, Manatee, and Gulf Coast Railroad, Peter B. Bradley, who created the American Agricultural Chemicals Company, secured the charter for the line in 1905 and renamed it the Charlotte Harbor and Northern Railway. His brother, Robert S. Bradley, became the railroad's president.

would buy tickets to travel to his resort and then pay large fees to stay at that resort. In honor of the festival, the resort was renamed the Gasparilla Inn.

Bradley chose a marketing stratagem that established a direct link between the resort and the pirate invasion. That link was of course Gasparilla himself. According to the stories circulating about Gasparilla, he actually lived on the island and supposedly buried his treasure there. Bradley knew that using Gasparilla for his publicity would be perfect. Tourists would be able to stay at the exact spot where Gasparilla lived, and they could even dig for his buried treasure along the beaches. The first step in this publicity campaign was to create a good backstory that appealed to tourists. The existing stories about Gasparilla seemed incomplete and lacked many of the bloodcurdling details that tourists expect. So, Bradly engaged a man named G. Pat LeMoyne to write a new story of Gasparilla that expanded on the current version. It was in LeMoyne's version that Gasparilla's name officially became José Gaspar.

LeMoyne did a great job. Most of the popular versions used in publications today came from LeMoyne's account. Among them is the meaning of his name Gasparilla, which LeMoyne explains as meaning "Gaspar the outlaw." LeMoyne also invented the sordid details about Gaspar's Spanish naval background and his theft of the Spanish crown jewels. He added details about Gaspar's base and described the finding of the skeletons of his victims in modern times. He created the story of the "little princess" and expanded on his association with Jean Laffite and Black Caesar. He also created the details of Gaspar's demise.

Needing a reliable foundation to give his version believability and legitimacy, LeMoyne named John Gomez as his primary source. He wrote, "This narrative was compiled by the writer from incidents told by John Gomez, better known as Panther Key John, a brother-in-law of Gasparilla and a member of his crew, who died at the age of 127 years at Panther Key, Fla., twelve miles below Marco, in 1900." LeMoyne created an imaginative backstory for Gomez and included him as a main character throughout the entire narrative, listing him as one of the ten survivors of Gaspar's crew. While it is true that John Gomez told tourists many stories about pirates and even told them that he had been a member of their crews, there is no evidence that he ever mentioned the name of José Gaspar or Gasparilla. I believe that LeMoyne used many of those stories told by Gomez and applied them to the version of the life of Gaspar that he was writing.

A brochure was published in 1913 that contained the story of José Gaspar and was complete with a frightening illustration of Gaspar on the cover. A highly effective advertisement for Gasparilla Inn, it was widely circulated

1913

A brochure advertising the Gasparilla Inn was produced and widely circulated. That brochure contained the story of José Gaspar.

throughout the community and everywhere the railway went. The brochure continued to be distributed in one form or another over the next ten years and perhaps longer. Unfortunately, there was a fire in the railway's records office in 1926 and none of the originals, or even copies, are known to have survived.

Until the release of the brochure in 1913, the full name of José Gaspar wasn't used in connection with the festival, at least not in print; Gasparilla was always used. That changed after the release of the brochure. His full name finally appeared in direct connection with the Gasparilla Festival on February 6, 1916, when *The Tampa Sunday Tribune* ran a story titled, *"Real Pirate Crew Will Come from Gasparilla"* and mentions that "old José Gaspar himself" will appear as the leader of the Gasparilla band.

The biggest boost for the legend of Gaspar came on February 1, 1917, when *The Tampa Daily News* published the entire story of his life, as written by LeMoyne. The article was titled, *"And Who Was José Gasparilla?"* and the source was credited in the byline as the C. H. & N. Railway. This was the version of Gaspar's life that has become the most popular and the most often recounted. Since it is believed that none of the original brochures have survived, this article provides the only source of the brochure's content. However, Gaspar might have remained among the fictional characters of legends and stories if not for the publication of *Pirates in the West Indies* in 1923.

Boston historian Francis B. C. Bradlee was in the process of writing a history of pirates in the West Indies. This was not intended as a work of fiction. Bradlee was writing a scholarly book and used news articles and naval reports as his primary sources. Francis was acquainted with another Bradley, although they spelled their names differently. The other Bradley was Robert S. Bradley, president of the Charlotte Harbor and Northern Railway. At some point, the railway executive sent a copy of his brochure on Gaspar to his friend the historian. By that time, a few new stories had been added, such as the story of Gaspar's planned attack of the steamboat, *Robert Fulton*. The Boston historian neglected to check any other sources and assumed the brochure was based upon historical fact. He included it, almost word for word, in his new book. Once in print by a reputable historian, the legend of Gaspar transformed into the realm of believability. In the eyes of many historians and enthusiasts, José Gaspar was now a real person and his background a matter of historical fact.

The "Cockeyed Lie"

Let's begin with the island of Gasparilla itself. According to the legend, the island of Boca Grande was named after Gaspar by Gaspar himself. In reality, the island was named after Friar Gaspar, a Spanish missionary who visited the Calusa in the 1600s. The name of Gasparilla appears on a 1774 map, many years before José Gaspar supposedly landed there. Additionally, the name "Gasparilla" means little Gaspar, not Gaspar the outlaw as stated in the legend.

Sanibel island was also named before Gaspar's arrival as seen on the same 1774 map. It was ether originally named "Santa Isybella" for Queen Isabella or was originally called Puerto de San Nivel. Either way, the island was known as Sanybel by the mid-18th century and as Sanibel by the 19th century.

Next. Let's take a close look at two of the main historical characters from the Spanish court who severely impacted Gaspar's life—Manuel Godoy and Maria Luísa. They were supposedly responsible for his exile from Spain. While they were indeed together in the Spanish court and had strong political influence, the years are wrong. In 1782 when Gaspar was in court, Godoy was 15 years old and Maria Luisa was actually married to the king's son, Carlos, future Carlos IV, King of Spain. It is extremely unlikely that she would have become romantically involved with a low-ranking member of the court. As for the mysterious little princess, she was said to be Josepha

Figure 38: *Boca Gasparilla and Sanybel Island shown on a 1774 Map*

de Mayorga, the daughter of Martín de Mayorga Ferrer, the Viceroy of New Spain. However, Martín died in 1782 and had no daughter. It is clear that those names were added to give the legend a sense of reality and legitimacy. But the person who inserted them into the legend failed to check the actual facts first.

Another interesting addition to the Gaspar legend was Gaspar's attempted capture of the steamboat, *Robert Fulton*. That story was apparently added to the brochure sometime between the 1917 news article and 1923, in time to be included in the historical book, *Pirates in the West Indies* written by Bradlee. Whereas Robert Fulton actually invented the steamboat, there was no steamboat named the *Robert Fulton* in the early 19th century. The U.S. Navy ordered a steam powered barge in 1814 that was originally named the *Demologos* but was renamed the *Fulton* by the time it was commissioned in June of 1815. However, the USS *Fulton* was more of a floating gun platform and could only travel short distances. Because of this, it never left New York Harbor. In 1821, the machinery was removed and replaced with sails. There was no steamship named *Robert Fulton* used in Cuban waters during the early 19th century.

As for Gaspar's pirate associates, Jean Laffite, and Black Caesar, they could not possibly have been with José Gaspar. Laffite's whereabouts are well documented after he left Galveston. In 1821 and 1822 when Jean Laffite was supposed to be sailing from Boca Grande with Gaspar, he was in reality either sailing the waters off the Bay of Campeche from his base on Mujeres or in prison in Cuba. And as we have already seen, Henri "Black" Caesar never really existed either. He was the product of local stories told to tourists in the early 20th century and of a 1921 work of fiction authored by Albert Payson Terhune titled *Black Caesar's Klan*.

The strongest case for the Gaspar legend being fictional came from one of the primary authors himself. G. Pat LeMoyne wrote the Gaspar story for the 1913 brochure promoting the Gasparilla Inn. He gave a history lecture in Fort Myers in 1949. In that lecture, LeMoyne stated that his story of José Gaspar was a "cockeyed lie without a true fact in it." He also said that he had written the brochure in a dramatic style that "tourists like to hear." He further explained that his story had actually been inspired by tall tales attributed to the colorful John Gómez, who claimed to have been a pirate and was known to sell fake treasure maps to "the gullible" for a "fancy figure."

LeMoyne's statements were refuted by E. D. Lambright, the original author of the Gasparilla legend as written for Ye Mystic Krewe of Gasparilla in 1904. An article was published in *The Tampa Tribune* on November 10, 1949, that stated, "E. D. Lambright, author of a book on Gasparilla, yesterday scoffed at claims that G. Pat LeMoyne, Sr. was originator of the José Gaspar pirate legend in 1913." It is interesting to note that E. D. Lambright was still the editorial director of *The Tampa Tribune* at the time that article was published. The article continued quoting Mr. Lambright as saying, "I have no doubt that Mr. LeMoyne had contributed to these legends, if he fabricated some for a publicity booklet. But to call the legend of José Gaspar a 'cockeyed lie' would seem to be controverted by other facts. The existence and operations of José Gaspar, the pirate, in Florida waters are fully authenticated by records in both Washington and Madrid."

Lambright's last statement brings us to the final argument for the non-existence of our intrepid pirate king. No such records exist, either in Washington or in Madrid. Dozens of scholarly researchers have examined thousands of Spanish records in their hopeful search for clues about the existence of José Gaspar, and so far, they haven't been able to find that name mentioned anywhere. As for records kept in Washington, D.C., one of the books I read suggested that in the early 1820s, Gaspar wrote a letter to the President of the United States, James Monroe, requesting amnesty in return for support against the enemies of the country. Just as with the research in Spain, there is no evidence that any such letter was ever written. Furthermore, there are no records or documents of any kind mentioning the name of José Gaspar in regard to a Florida pirate in the early 19th century anywhere in the naval archives or in any other records in the nation's capital.

The Gasparilla Legend Lives

In order to write a complete narrative on José Gaspar, it was necessary to dispel many of the assumed facts associated with Florida's pirate king. In doing so, it was not my intent to destroy Gaspar's legend but to simply separate that legend from the historical facts. I do not wish to take anything away from the millions of people who enthusiastically attend the Gasparilla Pirate Invasion or who enjoy portraying imaginary pirates. Truth and fantasy are two separate things, and it is my hope that one can learn the truth while still enjoying the fantasy.

The Gasparilla Pirate Invasion and other pirate festivals and events throughout the nation do a tremendous amount of good for their local communities. Festivals add an enormous boost to the local economies by bringing in thousands and sometimes millions of dollars. Ye Mystic Krewe of Gasparilla boasts that their organization is "dedicated to enriching the vitality and imagination of Tampa and the surrounding community." In my opinion, they are absolutely correct.

In addition to providing entertainment and bringing enjoyment to the community through their portrayal of pirates, organizations such as Ye

Mystic Krewe of Gasparilla as well as many other similar organizations are heavily involved in community outreach programs. They raise hundreds of thousands of dollars each year that go towards college scholarships as well as other charitable and community service programs. They regularly visit hospitals, assisted living facilities, and schools. They also host numerous children's events such as parades that get these young pirates involved in the community in a manner that they truly enjoy.

To quote the original newspaper article about the first Gasparilla Festival Ball which was published on May 10, 1904, *"Long live Gasparilla! Long may he reign!"*

The End

GLOSSARY

A

Acts of Union
Act passed by the British Parliament in 1707 that officially united England, Scotland, and Wales forming Great Britain.

Adams-Onís Treaty
treaty between the United States and Spain that was signed on February 22, 1819 and that ceded Florida to the U.S. and defined the boundary between the U.S. and Spanish territory.

Admiral
naval rank of a commander of a fleet or an officer of very high rank.

Aft
nautical term for the rear of a boat, ship, or any other type of vessel.

Amphibious Assault
an attack on land made from landing soldiers from a ship or vessel.

Anglicans
members of the Church of England, representing a middle ground between Catholicism and Protestantism.

Arawak Language
native American language originally spoken throughout the Caribbean and parts of northern South America.

Artillery Pieces
a class of heavy military weapons that encompasses a wide range of calibers which were built to fire projectiles at an enemy force during battle.

B

Baldric
belt for a sword or other piece of equipment worn over one shoulder and reaching down to the opposite hip.

Barcolongo
large Spanish fishing boat having two or three masts.

Bark
term used in the 18ᵗʰ Century for a nondescript vessel that did not fit any usual categories.

Bayou
a marshy outlet of a lake, bay, or river.

Below Decks
nautical term meaning the inside space of a ship or vessel which is literally below the main deck.

Bermuda Sloop
17ᵗʰ and 18ᵗʰ century single-masted vessel with a triangular sail rigged fore and aft with a square sail above that was designed for trade, measured approximately 70 feet in length, and had a large cargo hold.

Blunderbuss
short-barreled large-bored gun with a flared muzzle.

Boarding Action
when the crew of one vessel jumps over onto the second vessel during an attack.

Boatswain
senior member of the ship's crew who is in charge of operations on the deck.

Booty
valuable stolen goods. Synonyms: loot, plunder, haul, or pillage.

Boucan
Arawak word for the process of smoking meat over a slow burning fire.

Bow
front of a ship, boat, or vessel.

Bowline
knot used to form a fixed loop at the end of a rope but can also mean the anchor line.

Bowsprit
spar extending forward from a ship's bow, to which the forestays are fastened.

Brethren of the Coast
syndicate of pirate captains.

Brig

two-masted vessel with square rigging and an additional gaff sail on the mainmast.

Brigantine

two-masted vessel with a square-rigged foremast and a fore-and-aft rigged mainmast.

Broadside

the nearly simultaneous firing of all the guns from one side of a warship.

Buccaneer

17th century term meaning a pirate.

Calico Act

a series of English laws passed between 700 and 1721 which banned the import of most cotton textiles into England.

Calusa

a Native American tribe who originally inhabited the middle west coast of Florida from Tampa Bay south to the Everglades.

Calvinist Movement

a major branch of the Protestant Reformation that follows the theological teaching of John Calvin.

Captain General of the British Army

military rank for the senior general or the commander in chief of all generals.

Careening

the cleaning and repairing of the hull of a ship or vessel below the water line which requires taking it out of the water.

Carriage Gun

artillery piece that is mounted on a frame with small wheels, allowing the gun to be rolled short distances for loading and aiming.

Carronade

a smoothbore cast iron naval gun with a very short barrel, generally of a large caliber, that was used in close range battle against enemy ships or personnel.

Close the Distance

military term meaning to move toward and engage the enemy.

Colonel

military rank below a general for an officer who commands a regiment or a similar sized group of soldiers.

Colors
military term meaning a national flag. To raise the flag is to raise the colors. Colors flying is synonymous with flags waving.

Commissions
instruction, command, or authority given to individuals in writing.

Consort Vessel
small vessel used to carry supplies in support of other vessels.

Corsair
a pirate vessel usually associated with French pirates.

Creole
a stable natural language that develops from the simplifying and mixing of different languages into a new one, in the Caribbean it was generally African dialects mixed with either French or English.

D

Ducat
gold or silver trade coins with a fixed international value that were often used to conduct business between different nations.

Dutch East India Company
Dutch maritime trading corporation established 1602 which imported textiles, spices, and other valuable products from locations throughout the Indian Ocean.

Dutch Flute
16th through 18th century Dutch merchant ship with a rounded hull designed to carry large amounts of cargo.

Dutch Pinnace
fast and maneuverable 17th century Dutch ship with square rigging used for trade or as a small war ship.

E

Earthworks
an artificial fortification made of soil used for defense.

East Indiaman
any type of merchant vessel that belonged to an East India Company and carried goods between the Indian Ocean and Europe.

East Indies
the lands of India and South East Asia to include Malay Archipelago and Indonesia.

English East India Company

English maritime trading corporation established 1600 which imported textiles, spices, and other valuable products from locations throughout the Indian Ocean.

English Guinea

English gold coin minted between 1663 and 1813 and valued at one pound and one shilling.

English Lock or Doglock

a type of flintlock mechanism commonly used in England during the 17th century that ignites the gun powder in a firearm by striking a piece of flint against a metal frizzen causing a spark and has a small catch designed as a safety device to hold the cock in place.

English Pinnace

English term for a small service boat such as a row boat or long boat.

English Restoration

the restoration of the monarchy in England in 1660 when Parliamentary rule ended and Charles II was crowned king.

Felucca

a small vessel of Mediterranean design that has one mast rigged with a large triangular lateen sail.

Filibuster

17th century English term for pirate.

Fire Ship

any vessel intentionally set on fire and then sailed into an enemy fleet to cause panic.

Firing in Volley

military term meaning for all the soldiers in a designated group or line to fire their weapons at one time.

First Mate

senior member of the ship's crew who second in command to the quarter master.

Flag Ship

vessel used by the commanding officer of a group of naval ships.

Flintlock

type of lock mechanism introduced around 1550 that ignites the gun powder in a firearm by striking a piece of flint against a metal frizzen causing a spark.

Florida Straits

waterway that connects the Gulf of Mexico to the Atlantic Ocean and is located between Florida (including the Keys) and Cuba and the Bahama Islands.

Flyboat

Dutch designed small vessel with a flat bottom used as a costal cargo vessel or to service larger ships.

Fore Mast

mast in the front of the vessel usually in front of the main mast.

Forecastle

structure on the front of a ship or vessel usually containing all the ropes that are not in use.

French Arcadians

French colonists who settled in Arcadia (at the Gulf of St. Lawrence in Canada) during the 17th century.

French East India Company

French maritime trading corporation established 1664 which imported textiles, spices, and other valuable products from locations throughout the Indian Ocean.

Frigate

three-masted ship designed for war but is smaller in size and armament than a ship-o-the-line.

Fully Rigged

nautical term for a ship with three masts all of with have square sails.

G

Gaff Rigged

sail configuration plan for a four-sided sail rigged fore and aft and fastened at all four points with to a large spar connected to the center of the mast and hoisted up with lines connected to the top of the mast.

Galley

large vessel with up to three banks of oars that uses oars as its primary propulsion but also has sails.

Gaol

17th and 18th century term for a jail.

General

highest ranking military officer who is above the rank of Colonel and who commands several regiments or has supreme authority in military matters.

Glorious Revolution

revolution by the English Parliament in 1688 to overthrow King James II of England and place William III, Prince of Orange, who was James' nephew and son-in-law, and Mary II, who was James' daughter, on the throne.

Gunner

naval rank for a member of the crew who is an expert with naval ordinance and who is responsible for aiming and firing one or more of the ship's guns during battle.

Harbor Pilot

person who is thoroughly familiar with a harbor and guides large vessels into port.

Heretic

person holding a different religious belief than yours.

HMS

English nautical abbreviation meaning His/Her Majesty's Ships and precedes the name of all vessels officially listed as part of the Royal Navy.

Huguenots

16th and 17th century French Protestant denomination revived from the Calvinists.

Hull

main body of a ship or other vessel.

Hung from a Gibbet

the common practice of hanging the corps of a pirate or criminal in a metal cage in a public place as a deterrent to others.

Jacobite Revolution

a series of revolutions primarily conducted by the Scottish and Irish people in 1688, 1715, and 1745 to restore James II or his descendants to the throne of England.

Jamaica Sloop

17th and 18th century single-masted vessel with a triangular sail rigged fore and aft with a square sail above that was designed for trade, measured approximately 70 feet in length, but had a narrower construction than other sloops in increase speed.

Jolly Roger

a pirate flag which came from the French term "Jolie Rouge" which means "Happy Red".

K

Keel
the longitudinal structure along the centerline at the bottom of a vessel's hull, on which the rest of the hull is built.

Keys
small islands usually in a group such as the Florida Keys.

L

Larboard
the port, or left side of a ship, boat, or vessel.

Lateen Sail
large triangular sail originally designed by the Romans that gives a vessel more maneuverability than a square sail.

Leeward Passage
channel between Puerto Rico and the Virgin Islands by which vessels can exit the Caribbean in calmer waters.

Letter of Marque
document issued by an agent of a government giving the holder permission to attack all vessels of nations listed in the document which are deemed as enemies to the issuing government.

Line
nautical term meaning a working rope that is connected to a sail, anchor, or any part of a vessel used to sail it (a rope becomes a line as soon as it is attached to a part of the vessel).

Lock Mechanism
mechanism attached to the side of a firearm which is designed to ignite the gun powder in the barrel when a trigger is pulled.

Lord Protector of the Commonwealth of England
title given to the head of state of England during the time of Parliamentary rule known as the Commonwealth 1649–1660.

Louisiana Purchase
the acquisition of the territory of Louisiana by the United States from France in 1803.

M

Main Mast
tallest and most important mast on a ship or vessel.

Main Sail
on vessels rigged fore and aft, it is the largest sail and on vessels that are square rigged, it is the largest and lowest sail on the mast.

Major General
military rank for a general office below the rank of general.

Manila Galleon

Spanish merchant vessel that routinely traveled between the Philippines and the west coast of Mexico, primarily between Manila and Acapulco.

Man-O-War

class of warships in the 17th and 18th centuries that are the largest in size and armament.

Mark

an accounting value used in international trade that set one mark equal to two-thirds of one pound in weight of either silver or gold.

Marooned

the act of being set ashore and left alone on a deserted island. Usually done as a punishment for a serious violation of the ship's articles (rules).

Matchlock

larger version of the harquebus, it was introduced around 1520 and is a hand-held firearm that uses a match lock mechanism to ignite the gun powder.

Mate

member of a ship's crew.

Mediterranean Sea

a sea connected to the Atlantic Ocean, surrounded by the Mediterranean Basin and almost completely enclosed by land: on the north by Western and Southern Europe and Anatolia, on the south by North Africa, and on the east by the Levant.

Merchantmen

17th and 18th century term for any merchant ship or vessel.

Midshipman

naval rank for an officer cadet or someone who is in training to become a naval officer.

Militia

military force that is raised from the civilian population to supplement a regular military force.

Mikasuki

Native American tribe who originally inhabited Alabama, Georgia, and the panhandle of Florida.

Mint Julep

a drink which is common in the southeastern United States and is made with bourbon, mint, water, and sugar.

Mizzen Mast
mast directly behind the main mast.

Mogul
ruler of India or a member of the ruling dynasty of India, the Moguls were a warlike culture that concurred India in 1526 and remained in power until the 19th century.

Musket
a smoothbore firearm used by the military from the 16th century to the 19th century.

Musket Flint
a small purposely shaped piece of flint used to generate a spark on a flintlock and fire the musket.

Muskogee
Native American tribe commonly referred to a Creeks who originally inhabited Alabama, Georgia, and the panhandle of Florida.

Navigation Acts
series of English laws issued in the late 17th century that greatly restricted foreign trade, especially in the colonies and required that all foreign goods be shipped to the American colonies through English ports.

Nor'easter
storm or wind blowing from the northeast, usually accompanied by very heavy rain or snow.

Ordnance
military and naval term for artillery or guns onboard a ship.

Papist
person whose loyalties were to the Pope and the Roman Catholic Church, rather than the Church of England.

Pechelingues
16th and 17th century Spanish term for Dutch pirates.

Peninsular Campaign
campaign during the Napoleonic War that was fought between Brittan and France in Spain and Portugal.

Periauger
shallow draft, often flat-bottomed two-masted sailing vessel, often without a bowsprit, which also carried oars for rowing.

Pieces of Eight

term for a Spanish coin valued at eight reales but could also be applied to other Spanish real coins of lesser value.

Pilot

a ship's pilot was responsible for navigation and was a highly respected leadership position onboard all vessels.

Pink

three-masted, square-rigged sailing vessel, typically with a narrow, overhanging stern.

Pirate Banyon

gathering of pirate crews to socialize and form alliances, usually involving large quantities of food, drink, and general camaraderie.

Pirogue

a flat-bottomed canoe ideal for navigating backwater marshes and bayous.

Port

left side of a ship, boat, or vessel.

Pounds Sterling

denomination of English currency equal to 20 schillings.

Powder Flask

small flask containing gun powder and usually worn on a rope over the shoulder.

Presbyterians

members of the reformed tradition of Protestantism, influenced by the Reformed theology of John Calvin.

Pretender Across the Sea

18th century term for James Edward Stuart, the claimant to the throne of England who was often referred to as the "pretender" and who lived across the sea in France.

Privateers

private person or ship that engages in maritime warfare under a commission also known as letters of marque.

Prize Vessel

any vessel that has been captured by another vessel.

Protestant Reformation
1517–1648, a schism in Western Christianity initiated by Martin Luther and continued by John Calvin and other protestant reformers in 16th century Europe.

Pull Alongside
nautical term meaning to bring two vessels side by side.

Puritans
a member of a group of English Protestants of the late 16th and 17th centuries.

Quakers
member of the Religious Society of Friends, a Christian movement founded and devoted to peaceful principles, with a belief in the doctrine of "inner light" and rejection of formal ministry and all set forms of worship.

Quarter
military term meaning mercy shown toward an enemy or opponent, generally used as "No Quarter" or "Quarter Given".

Quarterdeck
raised deck behind the main or mizzen mast of a ship or vessel where navigation is done or where the captain commands the vessel.

Quartermaster
within a naval context of the 17th and 18th centuries, it was the rank of the senior member of the crew (not including the officers) who was responsible for the care and discipline of the rest of the crew.

Rating of Ships
classification of warships which indicated size and armament.

Real Situado
a Spanish term that referred to a year's amount of food, supplies, and money intended to feed, equip, and pay the salaries of a detachment of soldiers.

Re-provision
when used as a nautical term, it means to take on all manner of supplies on board a vessel.

Reformation of the Church of England
as incomplete and sought to simplify and regulate forms of worship.

Rigged Fore and Aft
configuration of sails on a vessel where triangular sails which are set along the keel.

Rigging

the system of ropes, cables, or chains to support a ship's masts (standard rigging) and to control or set the yards and sails (running rigging); and the action of providing a sailing ship with rigging.

Royal African Company

a company established by king Charles II of England in 1660 to trade along the West African coast for gold and slaves.

Royals

small sails flown immediately above the top gallants on square-rigged sailing vessels.

Rudder and Sails

nautical term meaning the navigational charts (rudder) and sails of a vessel, which was used as an expression "to take one's rudder and sails" meaning to keep one ashore.

S

Sailing Master

nautical rank for the officer responsible for setting and adjusting the sails of a ship or vessel.

Sailing Under False Colors

to raise the flag of another nation above one's ship or vessel.

Sale Before the Mast

term used by pirates for the auction of stolen goods for the crew after taking a vessel.

Schooner

two-masted vessel with sails that are rigged fore and aft.

Sea Beggars

17th century Dutch term meaning pirates.

Sea Rovers

17th century English term meaning pirates.

Seamen

general term for a person who serves aboard ship as a member of the crew.

Seaworthy

nautical term for a ship or vessel that is in good enough condition to sail on the sea.

Seersucker

thin and puckered cotton fabric that is commonly striped or checked.

Ship

three-masted vessel that is fully rigged with square sails.

Ship of the Line

largest and most powerful naval warships with either a first, second, or third rating usually carrying at least 80 guns.

Ship's Articles

rules for discipline and conduct agreed upon by the members of a pirate crew which could include anything from division of captured goods to treatment of prisoners.

Shot

military term for any type of projectile fired from a gun.

Sloop

single-masted vessel with a triangular sail rigged fore and aft.

Snow

large, two-masted merchant vessel that is rigged with square sails and can also be constructed with oars (galley-built snow).

South China Sea

body of water with Vietnam to the west, China to the north, the Philippines to the east, and Borneo to the south.

Spanish Doubloons

17th and 18th century English slag term for an extremely valuable Spanish gold coin.

Spanish Galleon

large Spanish merchant ship commonly used between the 16th and 18th centuries with three or four masts with the fore mast and the main mast rigged with square sails and the mizzen mast and aft mast rigged with lateen sails and a small square sail on a high-rising bowsprit.

Spanish Main

territory and waterways in the New World that was claimed by Spain in the 17th and 18th centuries that included the Caribbean, Gulf of Mexico, and the Bahamas.

Squadron

a small unit sent out from the main group to do some particular task, a naval squadron normally consists of between 5 and 15 vessels.

Square Rigged

vessel with traditional square sails that are generally perpendicular to the keel of the vessel.

Starboard
right side of a ship, boat, or vessel.

Steamship
a vessel powered by a steam engine turning one or two paddle wheels.

Stern
outer back rear of a ship, boat, or vessel.

Superstructure
any structure or cabin built on or above the main deck of a ship or vessel.

Sweeps
nautical slang term for the long oars used on any vessel.

Swivel Gun
small caliber gun that is mounted with a swivel device on the wall of a fort or on the rail of a vessel that can be easily aimed and fired at close range.

𝕿

Tainos
Native American group native to the Caribbean Islands.

Timucuan
Native American tribe that originally inhabited the central part of Florida from the east coast to the west coast.

Top Gallant
square sail immediately above the top sail on a square-rigged vessel.

Top Sail
square sail immediately above the main sail on a square-rigged vessel.

Tory
an American citizen who remained loyal to Great Brittan during the Revolutionary War.

Tradesmen
person who works in a trade such as a carpenter, black smith, gun smith, printer, etc.

Treaty of Breda
treaty between the Dutch Republic and England signed in 1667 that ended the Second Anglo-Dutch War that ceded the Dutch colonies in the mid-Atlantic to England.

Treaty of Madrid
treaty between England and Spain signed in 1670 and settled colonial disputes in America and officially ended the fifteen-year long war in the Caribbean that began with England's invasion and colonization of Jamaica.

Treaty of Osnabruck and Munster
treaty between most of the European nations signed in 1648 that ended the Thirty Years War.

Treaty of Paris
treaty between Brittan and France signed in 1763 that ended the Seven Years War.

Treaty of Ratisbon
treaty between Spain and France signed in 1684 that ended the War of the Reunions.

Treaty of Ryswick
treaty between France and the Grand Alliance signed in 1697 that ended the Nine Years War.

Treaty of Tordesillas
treaty between Spain and Portugal signed in 1494 that settled conflicts over lands newly discovered or explored by both Spanish and Portuguese explorers that divided the entire world in two halves, one belonging to Spain and the other belonging to Portugal.

Treaty of Utrecht
treaty signed in 1713 that ended the War of the Spanish Succession.

𝔘

Under Full Sail
nautical term meaning that all the sails are raised in order to sail as fast as possible.

Union Act of 1707
an act officially joining England and Scotland as one nation with one ruler.

𝔙

Vice Admiral
rank for a senior naval officer below the rank of admiral.

𝔚

Weather Gage
nautical term that means your vessel is up wind of the other vessel which gives you far more maneuverability that the vessel downwind.

West Indies
islands in the Caribbean that include the Greater Antilles, the Lesser Antilles, and the Lucayan Archipelago.

Windward Passage
the strait between Cuba and Hispaniola and extending along the Florida coast with consistently favorable winds blowing north by north east that was used as the primary route for vessels leaving the Caribbean.

BIBLIOGRAPHY

A Notorious Place, Florida, retrieved in October 2020 from http://pirates.hegewisch.net/fla.html, no copyright information given.

And Who Was José Gasparilla? Tampa Daily Times, 1 February 1917, p. 11. *April 5, Anniversary of the Casket Girl's Hunger Strike,* retrieved in October 2020 from http://www.medadvocates.org /celebrati/april/april_05.htm, no copyright information given.

Art History Resources. *18th Century Art and Baroque Art*, numerous works of art, retrieved from http://witcombe.sbc.edu/ARTH18thcentury.html, no copyright information given.

BLN—Boston Newsletter, 1704–1726, Boston, Massachusetts

Baumgarten, Linda. (1986) *Eighteenth Century Clothing at Williamsburg*, The Colonial Williamsburg Foundation, P.O. Box 1776 Williamsburg, VA.

Beater, Jack (2021) *Pirates and Buried Treasure of Florida,* Pineapple Press, Rowman & Littlefield Publishing Group, Inc., 4501 Forbes Boulevard, Suite 200, Lanham, Maryland 20706.

Bennett, Charles. (2001) *Laudonniere & Fort Caroline, History and Documents*, University of Alabama Press, Box 870380 Tuscaloosa, AL.

Biography of William Augustus Bowles, retrieved in October 2020 from https://accessgenealogy.com /georgia/biography-of-general-william-augustus-bowles.htm, no copyright information given.

Breverton, Terry. (2004) *Black Bart Roberts, The Greatest Pirate of Them All*, Pelican Publishing Company, Inc., 1000 Burmaster St, Gretna, Louisiana.

Breverton, Terry. (2004) *The Pirate Dictionary*, Pelican Publishing Company, Inc., 1000 Burmaster St, Gretna, Louisiana.

Brooks, Baylus C. (2016) *Quest for Blackbeard The True Story of Edward Thache and His World*, Lulu Press, Inc., Lake City, Florida.

Brown, Stephen R. (2003) *Scurvy, How a Surgeon, a Mariner, and a Gentleman solved the Greatest Medical Mystery of the Age of Sail*, Thomas Dunne Books, 175 5ᵗʰ Ave NY, NY 10010, An Imprint of St. Martin's Press.

Bradlee, Francis B. C. (1923) *Piracy in the West Indies and Its Suppression*, Newcomb & Gauss, Printers, Salem. Massachusetts.

Bruyneel, M. *Isle of Tortuga, Short History of Tortuga*, email: M.Bruyneel@ubvu.vu.nl, retrieved from http://zeerovery.nl/history/index.htm, copyright date not given.

Burgess, Robert & Clausen, Carl. (1982) *Florida's Golden Galleons, The Search for the 1715 Spanish Treasure Fleet,* Florida Classics Library, Port Salerno, FL.

CSPCS—*Calendar of State Papers, Colonial Series, America and the West Indies*, Preserved in the Public Record Office. Edited by Cecil Headlam. London: Cassell & Co. Ltd., 1930–1933.

Cahoon, Ben. *Chronology of Tortuga Governors, World Statesmen.org*, retrieved in October 2014 from http://www.worldstatesmen.org/Haiti.htm#Saint-Domingue, copyright date not given.

Canright, Marsha. (2015) *Jean Laffite: Pirate or Privateer?*, retrieved in May 2021 from https://www.coastmonthly.com/2015/09/jean-laffite-pirate-or-privateer.

Carrell, Toni L. PhD. *The U.S. Navy and the Anti-Piracy Patrol in the Caribbean*, retrieved in October 2020 from https://oceanexplorer.noaa.gov/explorations/08trouvadore/background/piracy/piracy.html, no copyright information given.

Causey, Donna R. *Biography: William Augustus Bowles Born 1763, a Young Tory*, retrieved in October 2020 from https://www.alabamapioneers.com/biography-william-augustus-bowles-born-1763/, no copyright information given.

Clifford, Barry & Kinkor, Kenneth. (2007) *The Real Pirates, the Untold Story of the Whydah*, National Geographic Society, Library of Congress Cataloging-in-publication 1145 17ᵗʰ St. NW Washington DC 20036-4688, Email: ngbookrights@ngs.org, National Geographic Books Subsidiary Rights.

Cooper, Luis. *Pirates in the Pensacola Bay Area: Fact or Fiction*? retrieved in October 2020 from https://www.visitpensacola.com/blog/post/pirates-in-the-pensacola-bay-area-fact-or-fiction, no copyright information given.

Cordingly, David. (2007) *Life Among the Pirates, the Romance and the Reality*, London, Little, Brown & Co.

Cordingly, David. (1996) *Pirates, Terror on the High Seas–From the Caribbean to the South China Sea*, London, Turner Publishing.

Cordingly, David. (2006) *Under the Black Flag, The Romance and Reality of Life Among the Pirates*, Random House Trade Paperback Edition.

Cordingly, David and Falconer, John. (1992) *Pirates Fact & Fiction*, Collins & Brown Ltd., London.

Cox, Dale. *Two Egg*, retrieved in October 2020 from https://www.twoeggfla.com/billybowlegs.html, no copyright information given.

Crime Museum. *Black Caesar*, retrieved in October 2020 from http://www.crimemuseum.org/crime-library/international-crimes/black-caesar, no copyright information given.

Dampier, William. (1968, 2007) *Memoirs of a Buccaneer, Dampier's New Voyage Round the World, 1697, First published by James Knapton, 1697*, Dove reprint of the work originally published in 1927 by Argonaut Press, London, under the title *A New Voyage Round the World* with introduction by Percy G. Addams, Dover Publications.

Davis, William C. (2006) *The Pirates Laffite, The Treacherous World of the Corsairs of the Gulf*, Harcourt, Inc., New Yok, First Harvest edition.

De Bry, John. A concise History of the 1715 Spanish Plate Fleet, 1715 Fleet Society, retrieved in October 2020 from https://1715fleetsociety.com/history, no copyright information given.

De Grummond, Jane Lucas. (1983) *Renato Beluche, Smuggler, Privateer, Patriot, 1780 - 1860*, Louisiana State University Press, Baton Rouge, Louisiana.

Eastman, Tamara J. and Bond, Constance. (2000) *The Pirate Trial of Anne Bonny and Mary Read*, edited by John Richard Stephens, Fern Canyon Press, P. O. Box 1708, Cambria, California.

Esquemeling, John. (2007) *The Buccaneers of America In the Original English Translation of 1684, First published by W. Crook, London, 1684*, Current edition published by Cosimo, Inc., Cosimo P.O. Box 416 Old Chelsea Station, NY, NY 10113, Email: www.cosimobooks.com.

Exquemelin, Alexander O. (2000) *De Americaensche Zee Roovers, First published by Jan ten Hoorn, Amsterdam, Holland, 1678, The Buccaneers of America*, Dover Publications, 31 East 2nd St. Mineola, NY 11501, translations Copyright 1969 by Alexis Brown, Introduction copyright 1969 by Jack Beeching, this edition.

Ford, Colleen. *Black Caesar: Florida's Most Mysterious Pirate*, retrieved in October 2020 from https://www.amli.com/blog/black-caesar-floridas-most-mysterious-pirate, no copyright information given.

Frethem, Deborah. (2013) *Haunted Tampa Spirits of the Bay*, The History Press.

Gerhard, Peter. (2003) *Pirates of New Spain 1575–1742*, Dover Publications, 31 East 2nd St. Mineola, NY 11501.

Hawkins, Paul. (2000, 2009) *Captain William Kidd*, Site design and layout Copyright by Paul Hawkins, pfrh@live.com, retrieved in March, 2014 from http://www.captainkidd.org.

Herriot, David. (1719) *The Information of David Herriot and Ignatius Pell contained the Appendix to: The Tryals of Major Stede Bonnet, and Other Pirates, London*, Printed for Benj. Cowse at the Rose and Crown in St. Paul's Church-Yard.

Historic Florida Militia. (2014) *Searle's Attack on Saint Augustine*, made possible in part by a grant from the St. Johns County Tourism and Development Council, retrieved in September 2014 from http://historic-florida-militia.org/events/annual/searles, Website by DeeLee Productions, LLC.

History of Pasco County Anclote, retrieved in October 2020 from http://www.fivay.org/anclote.html, no copyright information given.

History of The Gasparilla Inn & Club, retrieved in November 2020 from https://the-gasparilla-inn .com/about-us/history, no copyright information given.

Jean Lafitte: History & Mystery, retrieved in May 2021 from https://www.nps.gov/jela/learn /historyculture/jean-lafitte-history-mystery.htm, no copyright information given.

Jean Laffite: Mystery of the Unfound Treasure, retrieved in November 2020 from 2, no https:// pelicanstateofmind.com/louisiana-love/jean-lafitte-mystery-unfound-treasure, no copyright information given.

Johnson, Charles. (1999) *A General History of Pyrates*, Original edition published in London, 1724, second edition, 1724, third edition 1725, Daniel Defoe, A General History of the Pyrates, edited by Manuel Schonhorn. Dover Publication, Dover Publications Inc. 31 East 2nd St. Mineola, NY 11501.

Kaserman, James K. and Kaserman, Sarah. (2011) *Florida Pirates From the Southern Gulf Coast to the Keys and Beyond*, The History Press, Charleston, South Carolina.

Kazek, Kelly. *When French Orphans Called Casket Girls Came to Alabama as Wives for Colonists*, retrieved in October 2020 from https://www.al.com/living/2015/09/when_french_orphans _called_cas.html, no copyright information given.

Konstam, Angus. (1998) *Pirates 1660–1730*, Osprey Publishing, Osprey Direct, care of Random House Distribution Center, 400 Hahn Rd, Westminster, MD 21157.

Konstam, Angus. (2003) *The Pirate Ship 1660–1730*, Osprey Publishing, Osprey Direct, care of Random House Distribution Center, 400 Hahn Rd, Westminster, MD 21157.

Konstam, Angus with Kean, Roger Michael. (2007) *Pirates, Praetors of the Seas*, Skyhorse Publishing, and Text and Design, Thalamus Publishing, Skyhorse Publishing, 555 8th Ave., Suite 903, NY, NY 10018.

Konstam, Angus. (1999, 2002) *The History of Pirates*, The Lyons Press, PO Box 480, Guilford, CT 06437.

Konstam, Angus. (2007) *Scourge of the Seas, Buccaneers, Pirates, and Privateers*, Osprey Publishing, 443 Park Ave South, New York, NY 10016.

Kraus, Hans P. *Sir Francis Drake: A Pictorial Biography*, retrieved in October 2020 fromhttps://www .loc.gov/rr/rarebook/catalog/drake/drake-6-caribraid.html, no copyright information given.

Laflin, John Anderchyne. (2009) *The Journal of Jean Laffite*, as presented to the Missouri Historical Society, Moonglow Books, Columbia, South Carolina.

Landstrom, Bjorn. (1961) *The Ship, An Illustrated History*, Doubleday & Company Inc., Garden City, NY, also Produced by Interpublishing AB, Stockholm.

Little, Benerson. (2005) *The Sea Rover's Practice, Pirate Tactics and Techniques, 1630–1730*, Potomac Books Inc., 22841 Quicksilver Drive, Dulles, VA 20166.

Luenser, Don. *The Mystery of Jean Lafitte (Privateer, Smuggler, Pirate, Patriot, Spy) and His Lost Treasure*, retrieved in October 2020 from https://mysteriouswritings.com/the-mystery-of -jean-lafitte-privateer-smuggler-pirate-patriot-spy-and-his-lost-treasure-by-don-luenser, no copyright information given.

Louis-Michel Aury (1788-1821) July 6, 1818 Captain's Commission, Victor Decoville Estados Unidos de Buenos Ayres y Chile, DO4498, 18 June 2021, The Bryan Museum, Galveston, Texas.

Lyons, Douglas C. *The Legend of Black Caesar still haunts the Florida Keys*, retrieved in November 2020 from https://www.floridarambler.com/historic-florida-getaways/pirate-black-caesar -florida-keys, no copyright information given.

Mendes, Claudia. *The Legend of Black Caesar: The Pirate who Buried $6 Million Worth of Gold*, retrieved in November 2020 from https://www.warhistoryonline.com/instant-articles/the -legend-of-black-caesar.html, no copyright information given.

Marine Research Society. (1993) *The Pirates Own Book, Authentic Narratives of the Most Celebrated Sea Robbers*, originally published as Publication Number Four of the Marine Research Society, Salem, Massachusetts in 1924, Dover Publications, Inc., 31 East 2nd St. Mineola, NY 11501.

Millar, John Fitzhugh. *Buccaneers Davis, Wafer & Hingson, and the Ship Batchelors Delight*, William and Mary Alumni Association, One Alumni Drive, Post Office Box 2100, Williamsburg, VA 23187, retrieved in 2012 from https://www.wmalumni.com/?summer10_pirates, no copyright information given.

Nassau, Paradise Island Promotion Board. *Nassau, Our History, Tourism Internet Marketing by VERB*, retrieved from http://www.nassauparadiseisland.com/about-the-island/our-history, copyright 2014.

Parker, Anthony W. (1997) *Scottish Highlanders in Colonial Georgia*, The University of Georgia Press, Athens, Georgia.

Perez-Mallaina, Pablo E. (1998) *Spain's Men of the Sea, Daily life on the Indes Fleets in the Sixteenth Century*, Translated by Carla Rahn Phillips, Johns Hopkins University Press, Baltimore.

Petruzzello, Melissa. *Black Pirates and the Tale of Black Caesar*, Melissa Petruzzello is the Assistant Editor of Plant and Environmental Science Encyclopedia Britannica, Unity Wharf Mill Street,

London, SE1 2BH, England, retrieved in November 2020 from https://www.britannica.com /story/black-pirates-and-the-tale-of-black-caesar, no copyright information given.

Pine Island Information. *José Gaspar-Pirate*, retrieved in November 2020 from https://pineisland .info/index/island-legend/jose-gaspar-pirate, no copyright information given.

Pirate Black Caesar the Ex Slave, retrieved in November 2020 from https://www.pirateshipvallarta .com/blog/pirate-stories/pirate-black-caesar-ex-slave, no copyright information given.

Pirate Images, Pirate Maps, email: beej@beej.us, retrieved from http://beej.us/pirates/piratemaps .html, no copyright information given.

Powell, John T. (1998, 2004) *Military Artifacts from Spanish Florida, 1539 – 1821, An Internet Museum*, retrieved in 2010 from http://www.artifacts.org/default.htm.

Rare Gold Nuggets. *Jean Laffite, Jean Lafitte: Buried Treasures of the Notorious Pirate of the Gulf*, retrieved in November 2020 from http://raregoldnuggets.com/?p=5452, no copyright information given.

Reeves, Sally. *Searching for Laffite the Pirate*, retrieved in November 2020 from https://www .frenchquarter.com/jeanlaffitte/, no copyright information given.

See Cedar Key, retrieved in October 2020 from http://seecedarkey.com/history-of-cedar-key/, no copyright information given.

Seitz, Don Carlos. (2002) *Under the Black Flag: Exploits of the Most Notorious Pirates*, Courier Dover Publications.

Sewel Ford, Human Nature Author, Has the Clue to Gasparilla Gold. Tampa Daily Times, 6 April 1916, p. 3.

Sister, Susan. *51 Facts About Amelia Island*, retrieved in October 2020 from https://www .ameliaislandrealestate.net/posts/51-facts-about-amelia-island, no copyright information given.

The Colorful Pirate History of Amelia Island, retrieved in October 2020 from https://www.ameliavacations .com/the-colorful-pirate-history-of-amelia-island/, no copyright information given.

Spilman, Rick. *Panther John Gomez & the Legend of Gasparilla*, retrieved in June 2021 from http:// www.oldsaltblog.com/2018/01/panther-john-gomez-legend-gasparilla/.

The Famous Pirate Jean Lafitte, retrieved in November 2020 from http://www.famous-pirates.com /famous-pirates/jean-lafitte/, no copyright information given.

Treasure Expeditions. *Treasure of the Pirate José Gaspar, Also Known as Gasparilla*, retrieved in November 2020 from https://www.treasureexpeditions.com/pirate-treasure-jose-gaspar .html, no copyright information given.

Treasure Legends, retrieved in October 2020 from http://www.fivay.org/anclote.html, no copyright information given.

Troop, Alan F. *The Legends of Black Caesar*, retrieved in November 2020 from, https://www.sun-sentinel.com/news/fl-xpm-1991-10-06-9102090910-story.html, no copyright information given.

Vallar, Cindy. (2010) *Benjamin Hornigold, The Pirates' Pirate*, Cindy Vallar, Editor & Reviewer, P. O. Box 425, Keller, TX 76244, retrieved in January 2014 from http://www.cindyvallar.com/hornigold.html.

Vallar, Cindy. (2010) *Pirates and Privateers, The History of Maritime Piracy*, Cindy Vallar, Editor & Reviewer, P. O. Box 425, Keller, TX 76244, retrieved in September 2014 from http://www.cindyvallar.com/StAugustine.html#searle.

Visit St. Augustine, *Florida and the Pirates*, retrieved in October 2020 from https://www.visitstaugustine.com/history/castillo/castillo1.php, no copyright information given.

Warre, Harris Gaylord. *Laffite, Jean,* retrieved in November 2020 from https://tshaonline.org/handbook/online/articles/fla12 , no copyright information given.

Wombwell, James A. (2010) *The Long War Against Piracy: Historical Trends*, Combat Studies Institute Press, Fort Leavenworth, Kansas.

Wright, J. Leitch. (1960) *Andrew Ranson: Seventeenth Century Pirate? The Florida Historical Quarterly* vol. 39, no. 2, 1960, pp. 135–144. *JSTOR*, www.jstor.org/stable/30150254. Accessed 6 July 2021.2), 135-144. Retrieved July 6, 2021, from http://www.jstor.org/stable/30150254

Wright, J. Leitch Jr. (1967, 2010) *William Augustus Bowles Director General of the Creek Nation*, University of Georgia Press, Athens, Georgia.

INDEX

B

C

H

I

J

K

L

M

N

O

Ocracoke, North Carolina, 56, 58, 62–63, 68, 75–76

Old Providence Island, 118, 130, 132

Oriental, North Carolina, 63

P

Pamlico River, 63

Panama City, 27

Panther Key, 152, 155, 158

Panton, Leslie & Company, 83–84, 87

Patterson, Commodore Daniel, 99

Pauillac, Bordeaux, 91

Peninsular Campaign, 127, 174

Pensacola, 79–82, 87, 92, 97–98, 113, 115, 139, 153, 184

Pensacola, *Floridian* newsaper, 152–153

Peralta, Don Francisco, de 27

Philadelphia, 29, 53, 60–61, 63, 81, 111, 114, 131, 137

Pike, Zebulon, 106, 108

Pine Island, 76, 149, 188

Piques, Pedro, 24

Platt, Lieutenant, 142

Port-au-Prince, 89, 138

Porter, 143

Porter, Commodore David, 140, 142–43

Port Maria Bay, Jamaica, 71

Port Monroe, 141–42

Portobelo, 4–5, 21, 47, 50, 60

S

𝕿

𝖚

𝔙

W

Y

Z

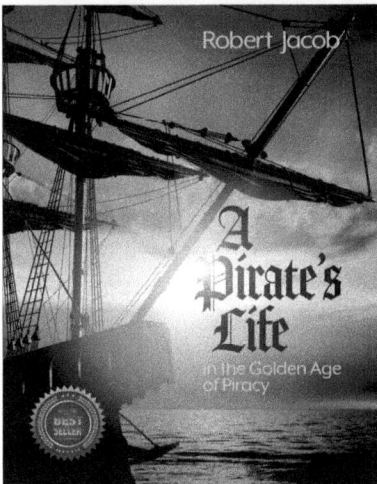

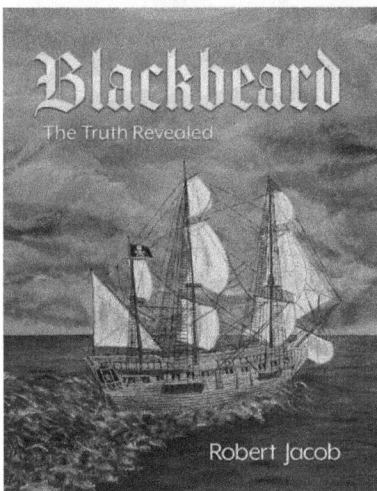

navigate you through the extensive array of records, including authoritative research and original eighteenth-century manuscripts, that corroborate the authenticity of Blackbeard's motivations, actions, and interactions with other pirates.